THE COMMONWEALTH AND INTERNATIONAL LIBRARY

Joint Chairmen of the Honorary Editorial Advisory Board

SIR ROBERT ROBINSON, O.M., F.R.S., LONDON

DEAN ATHELSTAN SPILHAUS, MINNESOTA

Publisher: ROBERT MAXWELL, M.C., M.P.

PHYSICS DIVISION

General Editor: W. ASHHURST

A Summary of A-Level Physics

A Summary
of A-Level Physics

BY

H. J. P. KEIGHLEY
Marlborough College

PERGAMON PRESS

OXFORD · LONDON · EDINBURGH · NEW YORK

TORONTO · SYDNEY · PARIS · BRAUNSCHWEIG

Pergamon Press Ltd., Headington Hill Hall, Oxford
4 & 5 Fitzroy Square, London W.1

Pergamon Press (Scotland) Ltd., 2 & 3 Teviot Place, Edinburgh 1

Pergamon Press Inc., Maxwell House, Fairview Park, Elmsford,
New York 10523

Pergamon of Canada Ltd., 207 Queen's Quay West, Toronto 1

Pergamon Press (Aust.) Pty. Ltd., 19a Boundary Street,
Rushcutters Bay, N.S.W. 2011, Australia

Pergamon Press S.A.R.L., 24 rue des Écoles, Paris 5ᵉ

Vieweg & Sohn GmbH, Burgplatz 1, Braunschweig

Printed in Great Britain by Bell and Bain Ltd., Glasgow

Preface to First Edition

This summary of A-level physics is based on duplicated notes which I have given to boys during their A-level course. It contains a summary of the facts and fundamental physical principles common to most A-level syllabuses and is intended for use in conjunction with a standard textbook. It can serve a number of purposes. Firstly, the sight of four rather large volumes covering the A-level syllabus may be rather forbidding, and to see it all summarized in a small volume can be a great incentive to learning. Secondly, many boys find it difficult to sort out the wood from the trees when reading a textbook, and a summary of the essential facts and principles which need to be mastered can be very helpful. Thirdly, a book which will fit into the pocket, containing a skeleton outline of the basic principles, is a useful aid when revising.

Many A-level syllabuses are very broad in their content, and, since it is not expected that candidates should cover the whole syllabus, a wide choice of questions is given in the examination. To cover all the topics of the various syllabuses would considerably increase the size of these notes and they would no longer serve the purposes for which they are designed. I have, therefore, tried to cover the topics which are common to nearly all the syllabuses and there are some blank pages at the back where additional notes on a few extra topics may be added.

I have omitted such topics as refractometers, measurement of coefficient of apparent expansion of a liquid and the method of mixtures (hardly a realistic method of specific heat measurement for present-day use) as they do not occur in some recently revised syllabuses. These omissions have allowed the inclusion of transistors and other more modern physics which are gradually being included in A-level syllabuses.

In view of the fact that the Ninth General Conference of Weights and Measures (1948) recommended that the use of calories should be discontinued, I have laid emphasis on the joule as the unit of heat but have included the calorie as some Examining Boards still use this rather obsolete unit.

In magnetism and electricity I have included both the " c.g.s. " and " m.k.s. " units.

I have tried to keep the use of calculus to a minimum in order to enable the book to be used by non-mathematical students.

It is inevitable that over the years the sources of many ideas gathered from colleagues and textbooks have been forgotten, but I acknowledge with gratitude all the help I have received from both.

Finally, my grateful thanks to Dr. F. R. McKim and Mr. W. G. Kalaugher who kindly read through the entire manuscript, and to Messrs. J. R. Mills, M. J. Harrison and S. W. Hockey, who read various parts of it. Also to Mr. J. R. Durling who was kind enough to read the proofs. I much appreciated their helpful criticisms and suggestions. I am also indebted to the publishers, especially Mr. W. Ashhurst, for all their help and advice.

Marlborough, 1966   H. J. P. K.

Preface to Second Edition

I HAVE included an additional chapter which includes topics in some syllabuses and not in others. These additions also mean that the book covers the syllabus for O.N.C. (basic).

A number of errors in the first edition have been corrected and a few alterations of a minor nature have also been made.

The standard S.I. abbreviations have been used throughout.

Marlborough, 1968   H. J. P. K.

How to Use this Book

THESE notes provide a summary of the essential facts and principles of A-level physics. You should use them in conjunction with a standard textbook. It is unlikely that you will understand the notes unless you have covered the work in class or have carefully read the appropriate section of a standard textbook. Once the work is *understood* the notes will provide a useful revision and an aid to committing the basic facts and principles to memory. The following suggestions may prove helpful.

1. By reading widely and by discussion, thoroughly master the basic principles.

2. Reproducing the diagrams and fundamental arguments on paper is the only way of making sure that you have thoroughly grasped the essentials. By doing this not only will you discover if the next step in the argument is understood but you will also find it easier to concentrate.

3. You should find that there are enough topics covered to enable you to have a reasonable choice of questions in any examination, but there are some blank pages at the back for you to write notes on additional topics. Remember that you will do better in an examination by knowing a fair proportion of the syllabus well than by having vague ideas about it all. It will help at revision time if you have previously gone through the notes and crossed out any sections which you have not studied.

4. If you follow the above suggestions, this summary should enable you to go through all the work and refresh your memory of the whole syllabus just prior to the examination.

5. No summary will enable you to do calculations. The only way to achieve this is to do a large number during your A-level course. Studying worked examples in a textbook may also prove helpful. Later try and work them without looking at the solution.

CHAPTER 1

Mechanics and the Properties of Matter

Fundamental Principles

Definitions

1. The *metre* is a certain number of wavelengths of light produced by a vapour lamp containing the element krypton.

2. The *kilogram* is the mass of a certain lump of platinum–iridium.

3. The *second* is a specific fraction of the year 1900. The Twelfth General Conference on Weights and Measures held in 1964 decided not to discard the astronomical definitions of the second, but to formalize the definition of an atomic second assigning a specific frequency to a transition between certain energy levels of the caesium-133 atom.

4. *Newton's laws of motion.*
  (i) If a body is at rest it will remain at rest, and if it is in motion it will continue to move with a constant velocity unless acted on by an external force.
 (ii) The rate of change of momentum of a body is proportional to the applied force and takes place in the direction of the applied force.
(iii) If a force acts on a body, then an equal and opposite force must act upon another body.

5. The *momentum* of a body at a given instant is defined as the product of its mass and its velocity. It is a vector quantity.

6. *Conservation of momentum.* If there are no external forces acting on a system of bodies in a certain direction, the total momentum of the system in that direction remains unchanged.

7. *Mass* is the quantity of matter in a body; it is inversely proportional to the acceleration when a given force acts on the body (usually measured in kilograms or grams).

8. *Force* is the product of mass and acceleration (usually measured in newtons or dynes).

9. A *dyne* gives a mass of 1 g an acceleration of 1 cm s^{-2} in the direction of the force.

10. A *newton* gives a mass of 1 kg an acceleration of 1 m s^{-2} in the direction of the force.

11. *Pressure* is force per unit area; the pressure at a point in a liquid is the limiting value of the ratio (force/area) for an area drawn about the point as the area goes to zero.

12. *Archimedes' principle.* When a body is totally or partially immersed in a fluid, it experiences an upthrust equal to the weight of fluid displaced.

13. *Law of flotation.* When a body floats it displaces its own weight of fluid.

14. The *weight* of a body on the earth is the force of attraction of the earth on it. The *kilogram-force* (kgf) is the force of attraction of the earth on a mass of 1 kg.

15. The *centre of gravity* of a body is the point through which its whole weight may be considered to act.

16. The *work* done by a force is the product of the force and the distance the point of application moves along the line of action.

17. An *erg* is the work done when a force of 1 dyne moves its point of application 1 cm in the direction of the force.

18. A *joule* is the work done when a force of 1 newton moves its point of application 1 m in the direction of the force. (1 joule $= 10^{7}$ ergs.)

19. *Energy* is the capacity for doing work (usually measured in ergs or joules).

20. *Kinetic energy* is the energy a body has by virtue of its motion (usually measured in joules or ergs).

21. *Potential energy* is the energy a body has by virtue of its position (usually measured in joules or ergs).

22. *Power* is the rate of doing work (usually measured in watts or h.p.).

23. A *watt* is a rate of working of 1 J s^{-1}.

24. *Conservation of energy.* Energy cannot be created or destroyed, but it can be converted from one form to another.

25. The *efficiency* of a machine is the ratio of the work got out to the work put in. (Multiplied by 100 it can be expressed as a percentage.)

26. The *mechanical advantage* of a machine is the ratio of the load to the effort.

27. The *velocity ratio* of a machine is the ratio of the distance moved by the effort to that moved by the load in the same time. (Efficiency = M.A./V.R.)

28. The *moment of a force* about a point is its turning effect about the point and is equal to the product of the force and the perpendicular distance from the point to its line of action.

29. A *couple* is a system of forces acting on a body, which produces only rotation. In its simplest form, it consists of two equal antiparallel forces not acting in the same straight line.

30. The *moment of a couple* is equal to the product of the force and the perpendicular distance between the forces (usually dyn cm or N m).

31. A *scalar* quantity is a quantity which has magnitude only and can be added algebraically, e.g. density, speed, energy.

32. A *vector* quantity has magnitude and direction, e.g. force, momentum, velocity and must be added by the Parallelogram law.

33. *Parallelogram law.* If two vectors are represented in magnitude and direction by the sides AB and AD of a parallelogram $ABCD$, then their resultant is represented in magnitude and direction by the diagonal AC.

34. *The laws of friction.*

(i) When the surfaces are at rest the frictional force may take any value up to a maximum known as the limiting friction F_1.

(ii) The limiting friction depends on the nature of the surfaces and is independent of the area of contact.

(iii) The coefficient of static friction μ_1 is defined as the ratio of the limiting friction F_1 to the normal reaction N_1 between the two surfaces, and is independent of the normal reaction. $F_1 = \mu_1 N_1$.

(iv) When sliding occurs, the frictional force, known as the sliding friction, is less than the limiting friction and is independent of speed. The coefficient of sliding friction μ_2 is defined as the ratio of the sliding friction F_2 to the normal reaction N_2 between the two surfaces. $F_2 = \mu_2 N_2$.

35. The *moment of inertia I* of a rigid body about a given axis is defined as Σmr^2, where m is the mass of each small element of the body, and r is the distance of the element from the axis of rotation. It is a measure of the opposition of a body to rotational acceleration.

36. *Conditions for equilibrium.*

(i) The resultant force on a body in any direction must be zero.

(ii) The sum of the clockwise moments about any axis must equal the sum of the anticlockwise moments about the same axis.

Uniformly Accelerated Motion

If an object undergoes a uniform acceleration a from an initial velocity u to a final velocity v in a time t, then by definition of acceleration

$$a = \frac{v-u}{t} \qquad \therefore v = u+at \qquad (1)$$

If the distance travelled is s, then $s = $ average velocity $\times$ time

$$\therefore s = \frac{u+v}{2}t = \frac{u+(u+at)}{2}t \quad \therefore s = ut+\tfrac{1}{2}at^2 \qquad (2)$$

Eliminating t from eqns. (1) and (2)

$$v^2 = u^2 + 2as \qquad (3)$$

Newton's second law of motion

Force $(F) \propto$ rate of change of momentum

$$F \propto \frac{mv - mu}{t} = \frac{m(v-u)}{t} = ma$$

From the definition of the dyne and the newton it follows that $F = ma$.

The law may be verified using the apparatus illustrated in Fig. 1. The rubber band is kept stretched a constant amount thus providing a constant force. The vibrating clapper makes dots on the

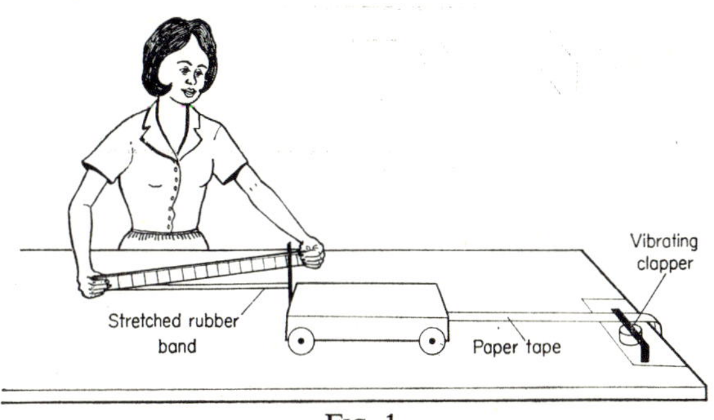

FIG. 1

paper tape at fixed intervals of time. The acceleration is calculated from the spacing of the dots. The force may be doubled, trebled and so on by using two or more rubber bands each stretched the same amount. The mass may be changed by adding bricks to the trolley. It is found that for a constant mass

$$\text{acceleration } (a) \propto \text{force } (F)$$

and that for a constant force

$$\text{acceleration } (a) \propto \frac{1}{\text{mass } (m)}$$

or $$F \propto ma \tag{4}$$

From the definition of a dyne or newton we have:

$$F \text{ (newtons)} = m \text{ (kg)} \times a \text{ (m s}^{-2})$$
$$F \text{ (dynes)} \quad = m \text{ (g)} \quad \times a \text{ (cm s}^{-2})$$

By Newton's third law of motion, whenever any two bodies in a system interact, the force on one is accompanied at every instant by an exactly equal and opposite force on the other. Hence the net force within the closed system is zero, and according to the second law there can be no change of momentum. This may be verified by compressing the springs in the two trolleys and releasing them. The two stops are adjusted so that the trolleys travel for the same time. Then, neglecting frictional forces, the distance s

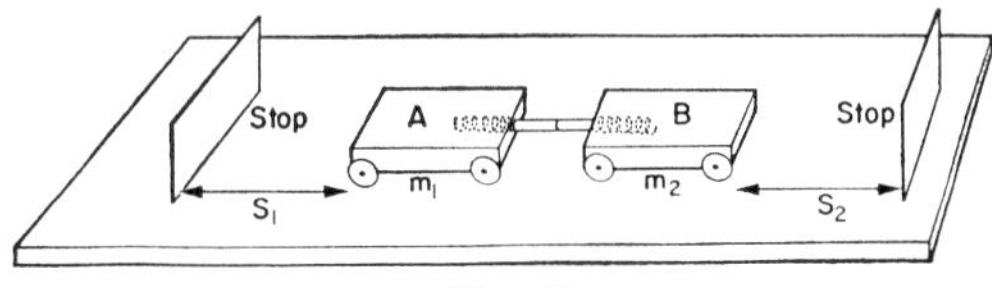

Fig. 2

travelled is proportional to the velocity. If m_1 and m_2 are the masses of the trolleys, we have to verify that $m_1 s_1 = m_2 s_2$.

Alternatively, a trolley, with a tape attached to the rear end running through a vibrator, and a pin on its front, is placed on a friction compensated slope. The trolley is pushed forward so that it collides with a stationary trolley and both trolleys move on together. The velocities before and after collision are calculated by analysis of the dots on the tape. The masses of the trolleys are known.

Sensitivity of a balance

The centre of gravity G is a distance d below the central knife edge F. Each arm of the balance is of length l. The weight of the balance is W and the weight of

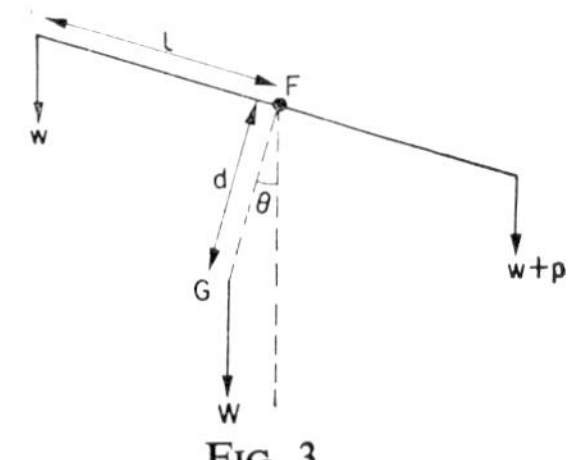

Fig. 3

each pan is w. If a weight p added to one pan results in a deflection θ, then taking moments about F:

$$wl \cos \theta + Wd \sin \theta = (w+p)\, l \cos \theta \qquad \therefore \ \tan \theta = \frac{pl}{Wd}$$

and if θ is small,

$$\text{sensitivity} = \frac{\theta}{p} = \frac{l}{Wd}$$

Kinetic energy (K.E.)

Suppose a mass m increases its velocity from u to v in a distance s. Then by eqn. (3), page 5, we have $a = (v^2 - u^2)/2s$. Using $F = ma$ we have:

$$Fs = \text{work done} = \tfrac{1}{2}mv^2 - \tfrac{1}{2}mu^2 = \text{change in kinetic energy}$$

Potential energy (P.E.)

If a mass m is raised through a height h, then P.E. $= mgh$.

Determination of the coefficients of friction

The laws of friction may be investigated using the apparatus illustrated in Fig. 4. To determine the coefficient of static friction weights are added to the pan until the block just starts to move.

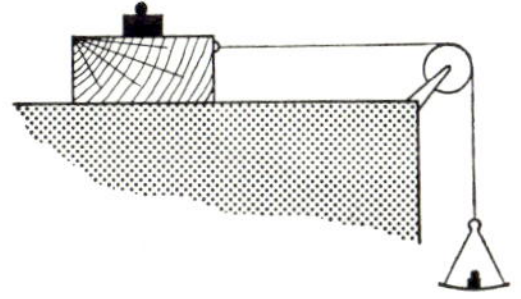

FIG. 4

The limiting friction F_1 is equal to the total weight of the scale pan and its load. The normal reaction N_1 is equal to the weight of the block plus the weight on top. The gradient of a graph of F_1 against N_1 is equal to μ_1. To determine μ_2 the load in the pan necessary to keep the block moving with a constant velocity is determined.

Circular Motion

Consider a body travelling in a circle of radius r with constant velocity v (Fig. 5). Suppose that it rotates through an angle θ in a time t. Angular velocity $\omega = \theta/t$ and $v = r\theta/t = r\omega$. By resolving the velocities into two directions at right angles, we have:

$$\text{Acceleration towards centre} = \frac{v \sin \theta - (-v \sin \theta)}{r \cdot 2\theta/v} = \frac{v^2}{r} = \omega^2 r$$

$$(5)$$

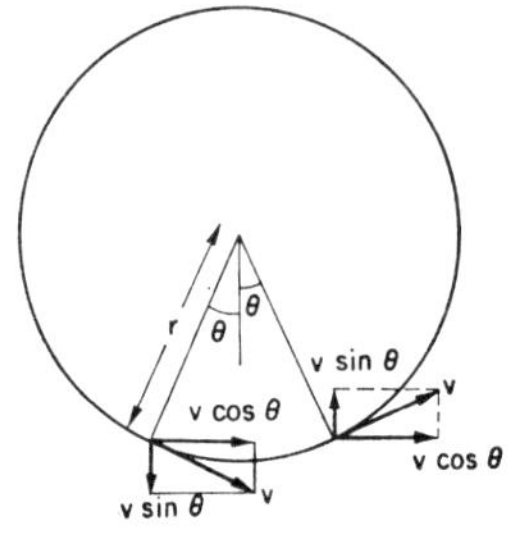

FIG. 5

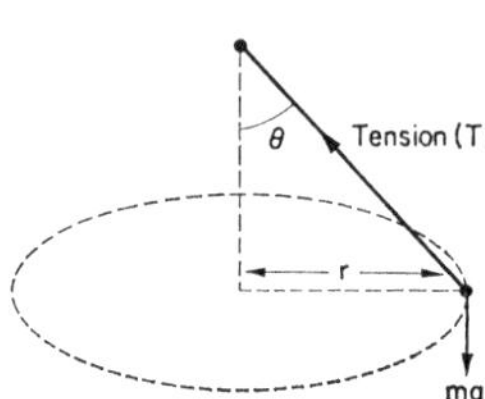

FIG. 6. Conical pendulum.

Conical pendulum (Fig. 6)

Principle of governor.

$$T \sin \theta = m\omega^2 r, \qquad T \cos \theta = mg$$
$$\therefore \ \tan \theta = \omega^2 r/g$$

Skidding and upsetting

Consider a car of mass m (Fig. 7) going round a left-hand bend. Let F_1 and F_2 be the frictional forces at the ground which prevent the car slipping, and R_1 and R_2 be the normal reactions at the wheels. Let the distance between the wheels be $2a$ and G be the centre of gravity. Then

$$F_1 + F_2 = \frac{mv^2}{r} \quad \text{and} \quad R_1 + R_2 = mg$$

Taking moments about G, $(F_1+F_2)h+R_1a-R_2a = 0$. From these equations it follows that

$$R_2 = \tfrac{1}{2}m(g+v^2h/ra) \quad \text{and} \quad R_1 = \tfrac{1}{2}m(g-v^2h/ra)$$

The car will overturn if $R_1\leqq0$, i.e. $v^2\geqq arg/h$. If μ is the co-efficient of friction then the maximum value of $F_1+F_2 = \mu mg$. The car will skid if $mv^2/r>\mu mg$.

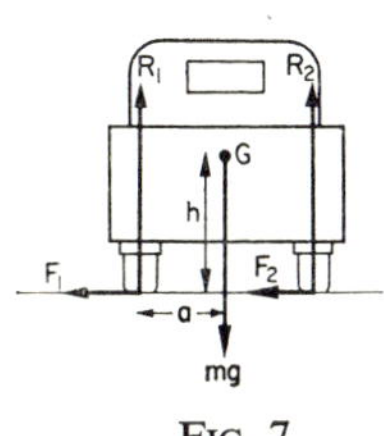

FIG. 7   FIG. 8

Banking

If there is no side slip, then the force towards the centre (Fig. 8),

$$(R_1+R_2) \sin \theta = mv^2/r.$$

For vertical equilibrium

$$(R_1+R_2) \cos \theta = mg.$$

$$\therefore \quad \tan \theta = v^2/rg.$$

Moment of Inertia (I)

Defined as Σmr^2 (see definition 35).

Routh's rule (applies to axis through centre of gravity for laminar or uniform solid).

$$I = \text{mass} \times \frac{\text{sum of squares of the two semi-axes}}{3 \text{ (rectangle, cube), } 4 \text{ (circle, ellipse), } 5 \text{ (sphere)}}$$

For a body rotating with angular velocity ω, couple $C = I(d\omega/dt)$, K.E. $= \tfrac{1}{2}I\omega^2$. For a rolling body the total K.E. $= \tfrac{1}{2}mv^2+\tfrac{1}{2}I\omega^2$.

Moment of Inertia of a flywheel (Fig. 9). The length of string is such that it becomes detached when the mass m reaches the ground travelling with velocity v after falling a distance h. By the conservation of energy

$$mgh = \tfrac{1}{2}mv^2 + \tfrac{1}{2}I\omega^2 + \text{work done against friction} \qquad (6)$$

If the wheel makes n revolutions after the string has become detached before it comes to rest in a time t, then average value of angular velocity $\omega = 2\pi n/t$. ω at instant string becomes detached

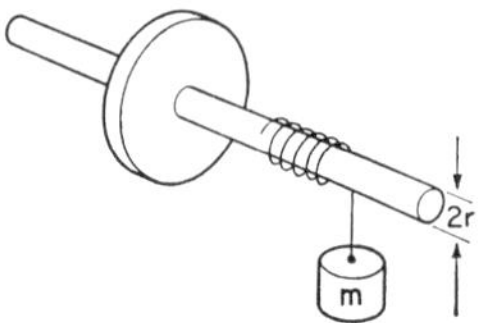

Fig. 9. Determination of moment of inertia.

is therefore $4\pi n/t$. Also $v = r\omega$. If F is the frictional couple, the work done against friction is $2\pi NF$, where N is the number of revolutions during the fall, and $2\pi nF = \tfrac{1}{2}I\omega^2$. Hence

$$2\pi NF = \frac{N}{n}\tfrac{1}{2}I\omega^2$$

and all the quantities in eqn. (6) are known except I which may be calculated.

Simple Harmonic Motion

Definitions

37. *Simple harmonic motion* is a motion in which the acceleration is proportional to the distance from a fixed point and is directed towards the point.

38. The *periodic time T* is the time for one complete oscillation.

39. The *amplitude a* is the maximum value of the displacement x from the centre of the oscillation.

Consider a particle P going round a circle of radius a with angular velocity ω.

Acceleration towards O (eqn. (5)) $= \omega^2 a$.

Acceleration of Q (the projection of P on AB)
$$= -\omega^2 a \cos(90° - \theta) = -\omega^2 a \sin\theta = -\omega^2 x$$

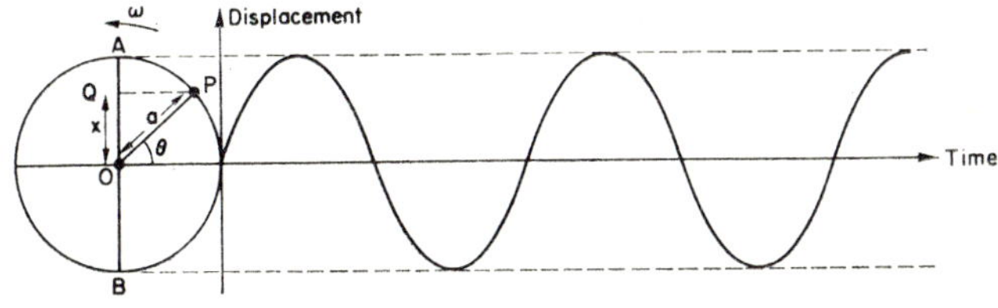

Fig. 10

i.e. the acceleration is proportional to the displacement and the motion is simple harmonic.

$$\text{Period } (T) = \frac{1}{\text{frequency } (f)} = \frac{2\pi}{\omega} = 2\pi \sqrt{\left(\frac{\text{displacement}}{\text{acceleration}}\right)}.$$

Velocity $v = a\omega \cos\theta$ and $\sin\theta = x/a$

But $\cos\theta = \sqrt{(1 - \sin^2\theta)}$ $\therefore$ $\cos\theta = \dfrac{1}{a}\sqrt{(a^2 - x^2)}$

$$\therefore v = \omega\sqrt{(a^2 - x^2)} \quad \text{and} \quad v_{\max} = a\omega.$$

To determine whether a motion is simple harmonic, displace the system and write down the restoring force for a displacement x from the equilibrium position. Apply Newton's second law of motion and if the resulting equation is of the form

$$\text{acceleration (acc.)} = -(\text{constant}) \times x$$

then the motion is simple harmonic with a period $T = 2\pi\sqrt{(x/\text{acc.})}$. The following examples illustrate the method.

Test-tube floating in a liquid (Fig. 11). When the tube of cross-sectional area A is displaced a distance x then by Archimedes' principle the restoring force $F = -Ax\rho g = ma$ (eqn.(4)). Therefore acc. $= -(Ag\rho/m)x$ (therefore s.h.m.) and

$$T = 2\pi\sqrt{(m/Ag\rho)}.$$

If h is the depth immersed when in equilibrium, then $m = hA\rho$ and

$$T = 2\pi\sqrt{(h/g)}.$$

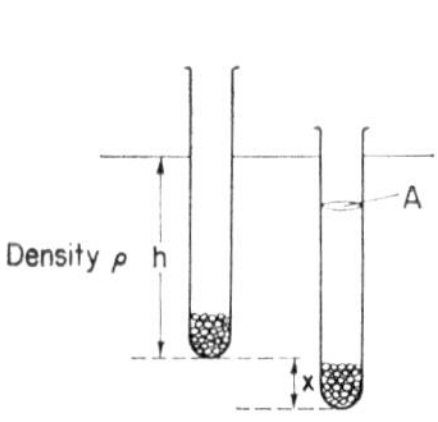

FIG. 11   FIG. 12

Liquid oscillating in a U-tube (Fig. 12). Restoring force when liquid is displaced x is $-2xA\rho g = ma = (Al\rho)a$. Therefore acc. $= -2xg/l$ (therefore s.h.m.) and $T = 2\pi\sqrt{(l/2g)}$.

Spiral spring (Fig. 13). If k is the force to extend the spring by unit length, then $Mg = ky$, where y is the extension for a load Mg. If the spring is now displaced a distance x from this equilibrium position, then restoring force

$$= -[k(y+x)-Mg] = Ma,$$

acc. $= -(k/M)x$ (therefore s.h.m.) and $T = 2\pi\sqrt{(M/k)} = 2\pi\sqrt{(y/g)}$ (using $Mg = ky$).

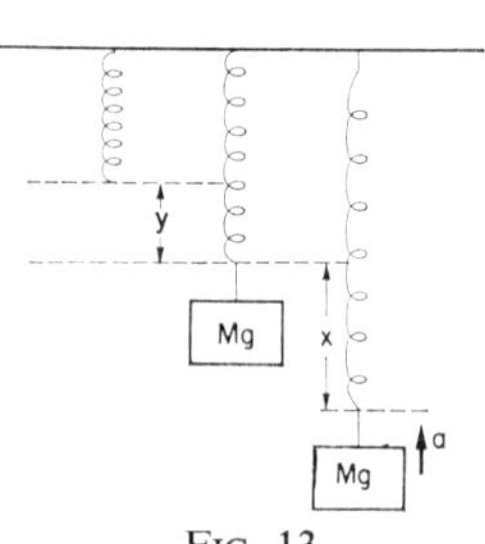

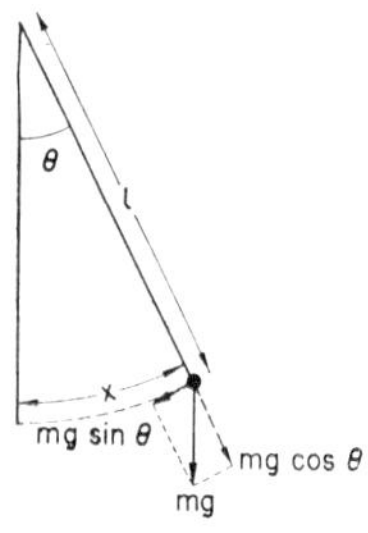

FIG. 13   FIG. 14

Simple pendulum (Fig. 14). For a displacement x, restoring force $= -mg \sin \theta = ma$. Therefore $a = -g \sin \theta = -g\theta$ (for small angles). Therefore acc. $= -(g/l)\,(l\theta)$ where $l\theta = x$ (therefore s.h.m.) and $T = 2\pi\sqrt{(l/g)}$.

Elasticity

Definitions

40. *Stress* is force per unit area (usually dyn cm^{-2} or N m^{-2}).

41. *Strain* is the fractional change in size or shape of a body when under stress, and has no units.

42. A wire is said to be *elastic* if it returns to its original length when the applied load is removed.

43. The *elastic limit* is the point beyond which the body does not return to its original dimensions when the load is removed.

44. The *yield point* is the point at which the material of the body begins to flow, the wire continuing to stretch for some time after the load has been reduced.

45. The *breaking stress* is the maximum stress that the specimen can sustain without fracturing.

46. *Hooke's law.* The extension of a wire is proportional to the applied load, provided that the extension is not too large.

47. The *limit of proportionality* is the point beyond which Hooke's law does not apply.

48. A modulus of elasticity is the ratio of stress/strain. In the case of stretched rods or wires this is called *Young's modulus*; the stress is the stretching force per unit area of cross-section and the strain is the increase in length per unit length. For fluids it is called the *bulk modulus*; the stress is the compressive force per unit area and the strain is the change in volume per unit volume. In the case of a shearing force it is called the *shear* or *rigidity modulus*; the stress is the tangential force per unit area and the strain is the angle of shear.

Load-extension graph for wires

Referring to Fig. 15, *A* is the *limit of proportionality*, *B* is the *elastic limit*, *C* is the *yield point*, *D* is the *breaking stress* and *E* the *point of fracture*. The exact shape of the curve depends on the material used.

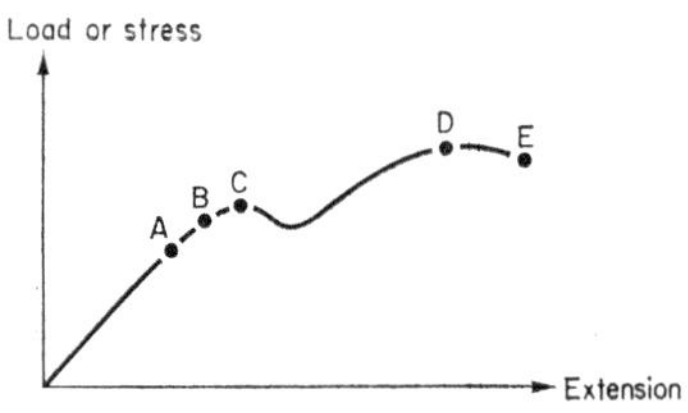

Fig. 15

Determination of Young's modulus

The vernier scale (see Fig. 16) is used to measure the extension *l* of the wire as the load *M* is increased. The readings are repeated as *M* is decreased. The length *L* of the wire from the support to the vernier is measured with a ruler. The diameter at several points is measured with a screw gauge and the cross-sectional area *A* is calculated. From the definition of Young's modulus we have

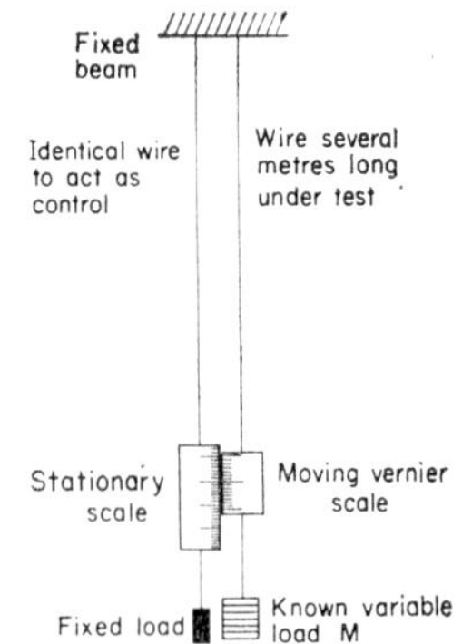

Fig. 16. Determination of Young's modulus.

$$\text{Young's modulus } E = \frac{\text{stress}}{\text{strain}} = \frac{Mg/A}{l/L}$$

A graph is plotted of *M* against *l* and

$$E = (\text{gradient of graph}) \times \frac{gL}{A}$$

By using two identical wires errors due to the yielding of the support and temperature changes are eliminated.

Energy stored in a stretched wire

The work done in increasing the extension by an amount x when the load is m is represented by the shaded area under the graph (Fig. 17). Hence the work done in extending the wire by L is the dotted area under the graph, i.e.

energy stored $=$ area under graph $= \frac{1}{2}$ final extension $\times$ final load

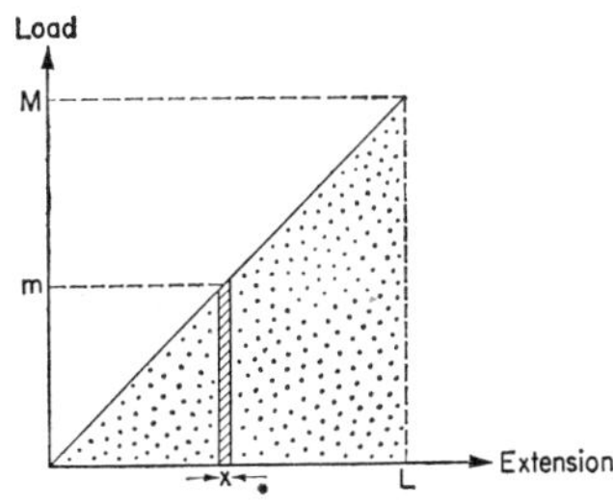

FIG. 17. Load-extension graph for a metal.

Force to hold a heated bar extended

Let F be the force needed to extend a wire of length L and cross-section A by l. Then by definition of Young's modulus $E = F/A \div l/L$. If α is the coefficient of expansion then for a rise in temperature $t°$, $l = L\alpha t$. Substituting for l we have

$$E = F/A \div L\alpha t/L \quad \text{or} \quad F = E\alpha t A$$

Gravitation

Definition

49. *Newton's law of gravitation.* Every particle of matter attracts every other particle of matter with a force which varies directly as the product of their masses m_1 and m_2 and inversely as the square of the distance r between their centres of mass, i.e.

$$F = G\frac{m_1 m_2}{r^2}$$

B

where G is the gravitational constant. (The law holds for rigid spherically symmetrical bodies provided we consider the whole mass of the body to be concentrated at its centre.)

Measurement of G. Boys's apparatus

The attractive forces between the gold and lead spheres (Fig. 18) result in a rotation θ. For accuracy the angle 2θ is measured, and the lead spheres are moved in each case so that the rotation is a

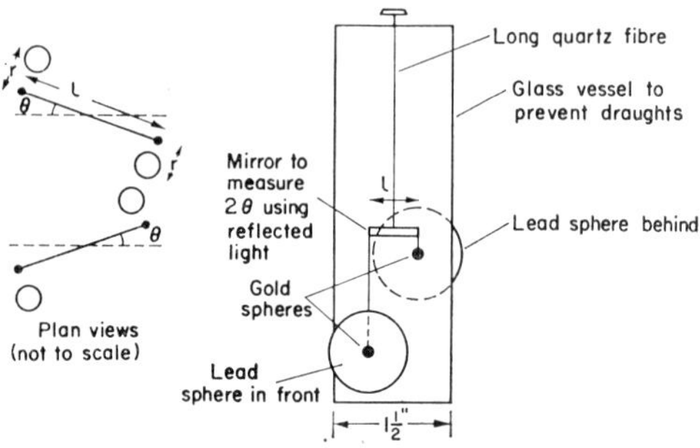

FIG. 18. Boys's method for *G*.

maximum. If m is mass of each gold sphere, M mass of each lead sphere, l the length of the mirror, r the distance between the centres of the spheres and c the couple to twist the suspension through an angle θ, then

$$c\theta = G\frac{mM}{r^2}l$$

The quantity c is found by a separate experiment. The suspended system is oscillated and the period T determined. Then if I is the moment of inertia of the suspended system $T = 2\pi\sqrt{(I/c)}$ and c may be calculated. The gold spheres are at different heights to reduce the effect of the attractions between the upper lead sphere and lower gold sphere and vice versa. The success of the experiment was due to the fact that the torsional constant c for quartz is very small although quartz has tremendous tensile strength. This enables the apparatus to be small so that it can be protected from draughts.

Determination of the mass of the Earth

For a mass m on the surface of the Earth $mg = G(mE/R^2)$, where E is the mass of the Earth and R its radius. Hence

$$E = gR^2/G.$$

Earth satellites

If a satellite of mass m travelling with velocity v is in a circular orbit of radius r, and if E is the mass of the Earth,

$$G\frac{Em}{r^2} = \frac{mv^2}{r} \qquad \therefore \ v^2r = GE = \text{constant}$$

Since v^2r is a constant it follows that for an orbit close to the Earth v must be large. The velocity required to keep a satellite in an orbit of radius r is independent of the mass of the satellite.

The motion of the Moon

Newton first tested his law of gravitation by observing the motion of the Moon. For simplicity of calculation we assume the Moon's orbit is circular. For a mass m on the Earth's surface we have already shown

$$m \times (\text{acceleration at Earth's surface}) = G\frac{mE}{R^2}$$

If r is radius of Moon's orbit and M is the mass of the Moon then

$$M \times (\text{acceleration of Moon}) = G\frac{ME}{r^2}$$

Hence

$$\frac{\text{acceleration of Moon}}{\text{acceleration at Earth's surface}} = \frac{R^2}{r^2} \qquad (7)$$

If T is the period of rotation of the Moon then its acceleration is $\omega^2 r = (2\pi/T)^2 r$. Substituting in eqn. (7), we have

$$\frac{(2\pi/T)^2}{g} r = \frac{R^2}{r^2}$$

Newton found that the two sides of this equation agreed to 1 part in 100.

Surface Tension

Definitions

50. *Surface tension* is the force per unit length acting on either side of a line drawn in the surface of the liquid. The direction of the force is perpendicular to the line and tangential to the surface (usually dyn cm^{-1} or N m^{-1}).

51. *Surface energy* is the work done against surface tension forces in enlarging the area of the surface by unity, the temperature being kept constant (usually ergs cm^{-2} or J m^{-2}).

52. The *angle of contact* is the angle, contained within the liquid, between the liquid surface and the solid surface.

Explanation of surface tension

Because of intermolecular forces a molecule in the surface of a liquid has a greater potential energy than a molecule in the body of the liquid. Like all mechanical systems, a liquid adjusts itself so that its total potential energy is a minimum. A liquid will therefore tend to have the smallest possible surface area. This accounts for the "taut" appearance of liquid surfaces and the fact that small drops of liquid tend to be spherical in shape.

Work done in increasing the area of a film

Suppose a soap film is formed on the wire frame (Fig. 19). Let the wire be moved a distance x increasing the area of each side of the film by A. The force on the wire due to each side of the film is Tl and thus the resultant force is $2Tl$. The work done is

$$2Tl \times x = T \times 2lx = T \text{ (increase in area of film)}.$$

Surface tension is therefore the work done in increasing the area of a film by unit amount.

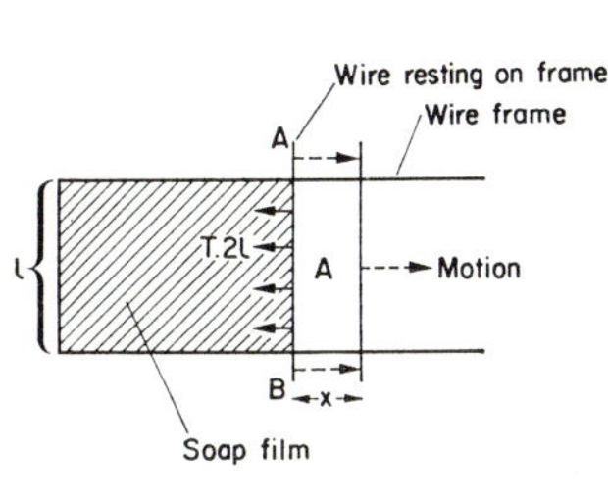

FIG. 19

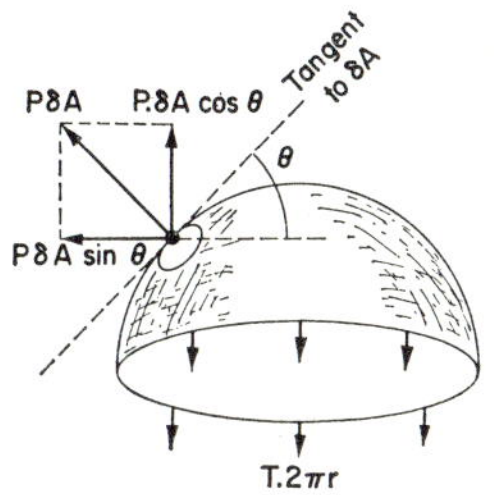

FIG. 20

Excess pressure inside a bubble

We consider the equilibrium of half the bubble. The surface tension force T acts all round the circumference of the base of the hemisphere (Fig. 20). The resultant force on the base is $T \cdot 2\pi r$. This force is balanced by the components perpendicular to the base of the forces acting on the curved surface due to the excess pressure in the bubble. Consider an element of area δA. The force on this is $p\delta A$.

$$\Sigma p\delta A \sin \theta = 0 \quad \text{and} \quad \Sigma p\delta A \cos \theta = T \cdot 2\pi r$$

But $\Sigma \delta A \cos \theta = \pi r^2$ since $\delta A \cos \theta$ is the projection of δA on the base of the hemisphere. Hence

$$p \cdot \pi r^2 = T \cdot 2\pi r \quad \text{or} \quad p = \frac{2T}{r}$$

For a soap film which has two sides, $p = 4T/r$.

In general, on the concave side of any spherical liquid surface, the pressure is greater than that on the convex side by $2T/r$. In Fig. 21 the smaller bubble shrinks when the taps T_2 and T_3 are opened because the pressure inside the smaller bubble is greater than the pressure inside the larger bubble.

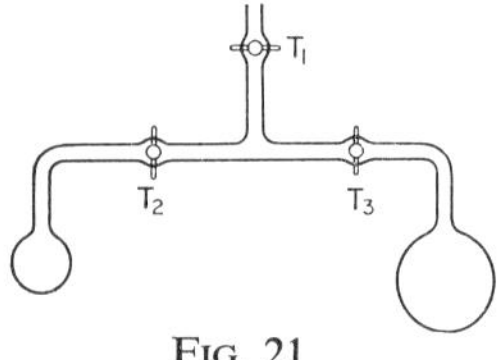

FIG. 21

Why floating bodies attract each other (Fig. 22)

When the two bodies are close together the liquid rises in between them because of capillary attraction. The pressure on the convex side of the curved surface is $(A - 2T/r)$, where A is the atmospheric pressure, r the radius of curvature of the surface of the

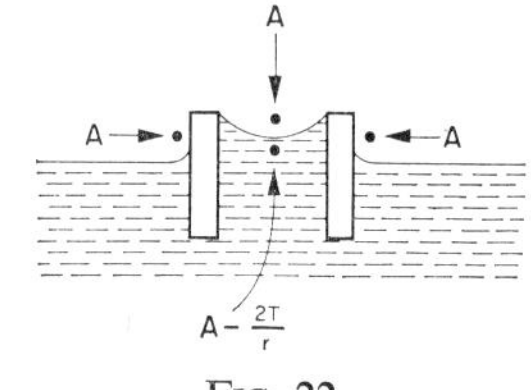

FIG. 22

liquid between the objects and T the surface tension. The pressure between the objects is thus reduced and the objects move towards each other.

Measurement of surface tension

(i) *Capillary rise* (Fig. 23). Let R be radius of meniscus and r be radius of tube. If the liquid of surface tension T and density ρ has an angle of contact θ, then pressure at $X = \text{atm} - 2T/R + \rho gh = $ pressure at $Y = \text{atm}$ (where atm = atmospheric pressure).

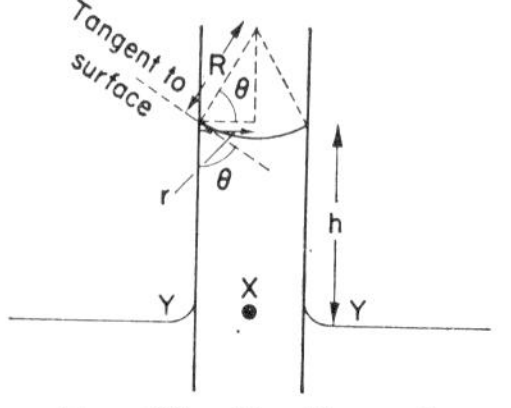

FIG. 23. Capillary rise,

But $r = R \cos \theta$ and hence $2T/(r/\cos \theta) = \rho gh$

or $$T = \rho ghr/2 \cos \theta$$

h is conveniently measured using a pair of dividers, and r (at the meniscus) by breaking the tube and using a travelling microscope. It is very important to ensure that the tube is clean.

(ii) *Using a chemical balance.* The glass plate (Fig. 24) is adjusted so that its lower surface is in the surface of the liquid when the balance beam is horizontal. The additional mass m necessary to pull the plate out of the liquid is determined. If a is the thickness of the plate and b is its length, then

$$T(2a+2b) = mg$$

and hence T may be determined.

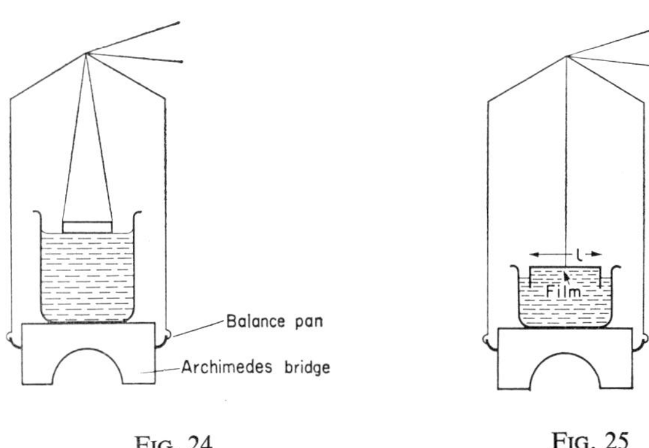

Fig. 24

Fig. 25

In the case of a soap film a wire frame is used instead of the plate. The beam is balanced and the film punctured. If weights m are removed until the beam is again horizontal, then if l is the length of the frame (Fig. 25)

$$mg = 2Tl$$

and hence T may be determined.

Viscosity

Definitions

53. The property of a fluid which tends to prevent one layer moving over another is called *viscosity*.

54. For many fluids, if the flow is streamline the tangential stress on any layer is proportional to the velocity gradient (at right angles to the direction of flow) in that layer. The constant of proportionality is called the *coefficient of viscosity η* (usually N s m^{-2} or dyn s cm^{-2}).

55. The velocity at which the flow becomes turbulent is known as the *critical velocity*.

Poiseuille's formula for flow of a viscous liquid through a capillary tube

The following derivation illustrates the method of dimensions. Suppose the rate of flow V depends on the radius of the tube r, the pressure gradient along the tube (p/l) and the coefficient of viscosity η, then

$$V = k\eta^a r^b (p/l)^c$$
$$\therefore \ L^3 T^{-1} = k (ML^{-1}T^{-1})^a L^b (ML^{-2}T^{-2})^c$$

Equating dimensions we have

$$0 = a+c, \quad 3 = -a+b-2c, \quad -1 = -a-2c$$

whence $a = -1$, $b = 4$, $c = 1$.

$$V = k\frac{pr^4}{l\eta}$$

If the flow is slow enough to neglect the work done in giving the liquid kinetic energy, then experiment shows that $k = \pi/8$.

Stokes's law

If a sphere of radius r is moving through a large expanse of fluid of coefficient of viscosity η, and velocity v, then the viscous drag is given by

$$F = 6\pi\eta rv$$

If the viscous medium is a liquid of density σ, the material of the sphere of density ρ, and the terminal velocity of the sphere is

v, then

$$\tfrac{4}{3}\pi r^3 \rho g = 6\pi\eta rv + \tfrac{4}{3}\pi r^3 \sigma g$$

$$\therefore \; \eta = \frac{2}{9}\frac{gr^2}{v}(\rho-\sigma)$$

and hence η may be determined.

Variation of viscosity with temperature

The viscosity of most liquids is probably due to temporary bonds between adjacent molecules. When the temperature is increased, the thermal motion will shorten the duration of the bonds and the viscosity will decrease. With gases η increases with temperature increase. When the temperature is high more molecules are likely to cross a layer every second. The rate of transfer of momentum and hence the viscous force will be greater.

CHAPTER 2

Heat

Expansion

Definitions

56. The *coefficient of linear expansion* of a substance is the increase in length of unit length of the substance at $0°C$, for a degree rise in temperature (usually $degC^{-1}$).

57. The *coefficient of cubical expansion* of a solid, liquid or gas at constant pressure is the increase in volume of unit volume at $0°C$ for a degree rise in temperature (usually $degC^{-1}$).

*Laboratory method for the coefficient of expansion
of a solid*

The apparatus illustrated in Fig. 26 may be used. Procedure: Measure length of rod with a ruler (L centimetres). Read screw gauge (x_1 millimetres), unscrew. Read temperature ($t_1°C$). Pass steam until screw gauge reading is constant (x_2 millimetres) when taken at fixed intervals of time. Read final temperature ($t_2°C$). Then

$$\text{coefficient of expansion } \alpha = \frac{(x_2 - x_1)/10}{L(t_2 - t_1)} \text{ per } °C$$

*The relationship between the coefficient of expansion
of a liquid and density*

If a liquid has a volume V_0 at $0°C$ and V_t at $t°C$, and if γ is the coefficient of cubical expansion $V_t = V_0(1 + \gamma t)$. Let M be the

24

mass of liquid at both temperatures, ρ_0 the density at 0°C, and ρ_t is the density at t°C then $V_0 = M/\rho_0$ and $V_t = M/\rho_t$. Hence

$$M/\rho_t = M/\rho_0(1+\gamma t) \quad \text{or} \quad \rho_0 = \rho_t(1+\gamma t) \qquad (8)$$

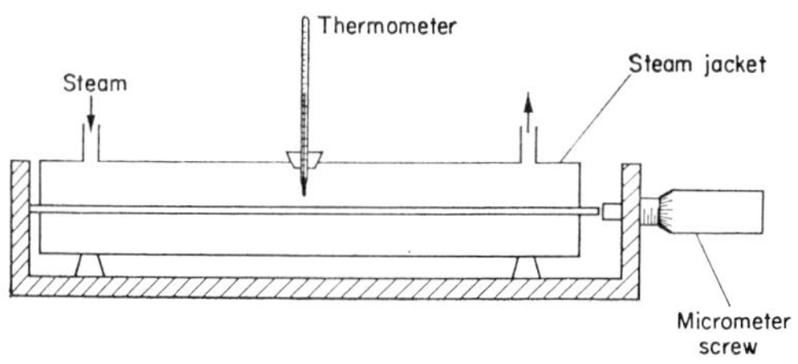

Fig. 26

Real coefficient of expansion of a liquid

The apparatus illustrated in Fig. 27 is very convenient. The heights h_1 and h_2 are measured with a cathetometer. Let the pressure of air above the liquid be p and let ρ_1 be the density at

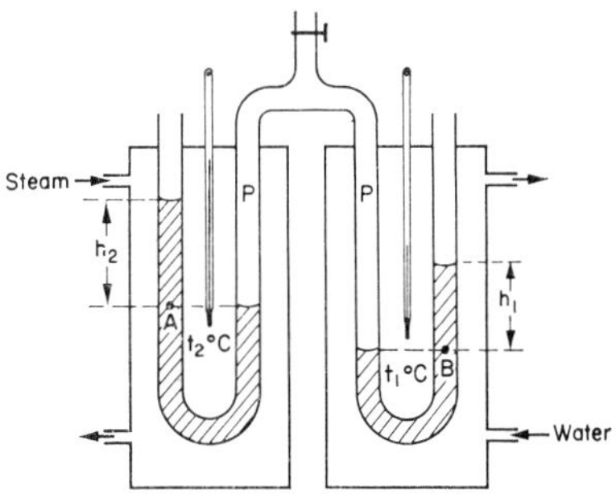

Fig. 27

t_1°C and ρ_2 be the density at t_2°C. If atm represents the atmospheric pressure, then

$$\text{pressure at } A = \text{pressure at } B = p$$

$$\text{atm} + \rho_2 g h_2 = \text{atm} + \rho_1 g h_1 \qquad \therefore \ \rho_2/\rho_1 = h_1/h_2 \qquad (9)$$

Using eqn. (8) we have $\rho_0/\rho_1 = (1+\gamma t_1)$, and $\rho_0/\rho_2 = (1+\gamma t_2)$. Substituting in eqn. (9)

$$\frac{1+\gamma t_1}{1+\gamma t_2} = \frac{h_1}{h_2}$$

$$\therefore \; \gamma = \frac{h_1 - h_2}{h_2 t_1 - h_1 t_2}$$

Specific Heat and Latent Heat

Definitions

58. The unit of heat energy is the *joule* (J) equal to 10^7 ergs. The 15°C *calorie*, defined as the heat required to raise the temperature of 1 g of water from 14·5°C to 15·5°C, is the specific heat of water at 15°C and is equal to 4·1855 J.

59. The *British Thermal Unit* or Btu is the quantity of heat energy required to raise the temperature of 1 lb of water through 1 degF (1 therm $= 10^5$ Btu).

60. The *thermal capacity* of a body is the quantity of heat energy required to raise its temperature through 1 degree (usually $J\,degC^{-1}$ or $cal\,degC^{-1}$).

61. The *water equivalent* of a body is the mass of water having the same thermal capacity as the body itself (usually kg or g).

62. If a quantity of heat Q is supplied to a mass m of a substance thereby raising its temperature from T_1 to T_2, then the *mean specific heat* over the temperature range T_1 to T_2 is given by

$$s = \frac{Q}{m(T_2 - T_1)}$$

As T_1 approaches T_2 the quantity s approaches a limit known as the *true specific heat* at that particular temperature (usually $J\,kg^{-1}\,degC^{-1}$ or $cal\,g^{-1}\,degC^{-1}$).

63. The *latent heat of fusion* is the quantity of heat energy required to change unit mass of solid to liquid without change of temperature (usually $J\,kg^{-1}$ or $cal\,g^{-1}$).

64. The *latent heat of vaporization* is the quantity of heat energy required to change unit mass of liquid to vapour without change of temperature (usually $J\,kg^{-1}$ or $cal\,g^{-1}$).

65. The *calorific value* of a food or fuel is the quantity of heat given out by unit mass when it is completely burnt.

Specific heat of a solid

A coil of wire is wound on a cylinder of the substance. There must be good thermal contact between the coil and the cylinder, and the cylinder is lagged. A thermometer is inserted in a hole in the cylinder. If a p.d. of V volts maintains a current of I amps through the coil for t seconds then neglecting heat losses

$$V.I.t = m.s.\theta$$

where m is the mass of the cylinder in grams, θ the temperature rise in t seconds and s the specific heat of the substance in joules per gram per degree C. In accurate work the cylinder is suspended in a vacuum.

Specific heat of a liquid

A heating coil is immersed in a liquid, the liquid being contained in a lagged calorimeter (Fig. 28). Suppose a p.d. of V volts maintains a current of I amps for t seconds. Let m grams be mass of liquid of specific heat s joules per gram per degree C in a calorimeter of water equivalent W grams and let θ be the rise in temperature. Then

$$VIt = (ms + Ws')\theta$$

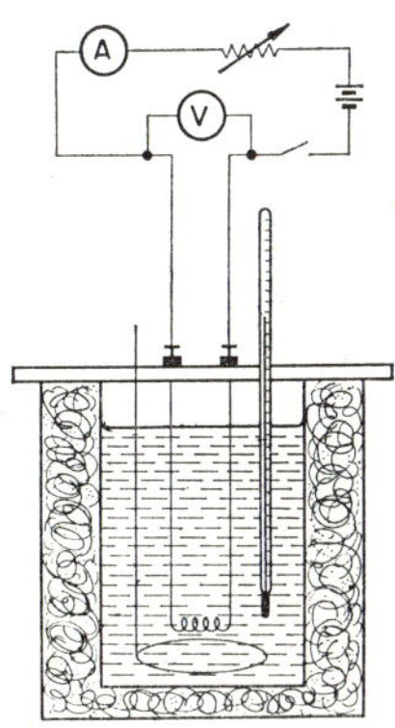

FIG. 28

where s' is the specific heat of water and everything in the equation except s is known. In order to reduce the error due to heat exchanges with the atmosphere start with the liquid a few degrees below room temperature and finish the same number of degrees above.

If water is the liquid in the calorimeter the experiment may be used to determine the electrical (mechanical) equivalent of heat J. In this case $VIt/J = (m+W)\theta$.

Continuous flow calorimetry

The apparatus illustrated in Fig. 29 is sensitive enough to detect the variation of the specific heat of water with temperature and was designed by Callendar and Barnes. Water from a

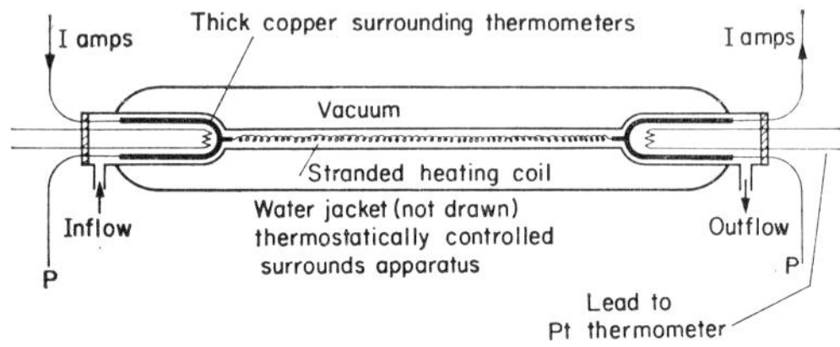

FIG. 29. Callendar and Barnes's continuous flow apparatus.

constant head apparatus flows through the central tube. The p.d. V volts is measured by connecting the leads PP into a potentiometer circuit. The current I amps is also measured using a potentiometer. The thick copper surrounding the thermometers serves to equalize the temperature and prevent the generation of heat in the vicinity of the thermometers. The platinum thermometer pair are connected differentially. If m grams of water flow per second and the rise in temperature of the water is θ degC then

$$VI = ms\theta + h$$

where s is the specific heat of water in joules per gram per degree C and h is the heat lost to the surroundings. h may be eliminated by conducting a second experiment with a different rate of flow m_1, the wattage being adjusted by trial and error to give the same temperature rise as in the first experiment. Then $V_1 I_1 = m_1 s\theta + h$. Subtracting these equations we get

$$VI - V_1 I_1 = (m - m_1)s\theta$$

and hence s may be determined. s is, of course, the number of joules per calorie and is numerically equal to the mechanical equivalent of heat.

The advantages of continuous flow calorimetry are (i) the conditions are steady when the temperatures are read and the thermometer readings may be taken very accurately, (ii) the thermal capacity of the apparatus is eliminated, and (iii) the heat losses are small and they may be eliminated.

Callendar's rotating drum (Fig. 30)

Let W be weight on the bottom, T the reading on the spring balance, n the number of revolutions, θ the temperature rise, M the mass of the drum of material of specific heat s, M_1 the mass of water in the drum of specific heat s_1, r the radius of the drum and F the frictional force. Then equating the work done against the frictional couple to the heat energy produced we have:

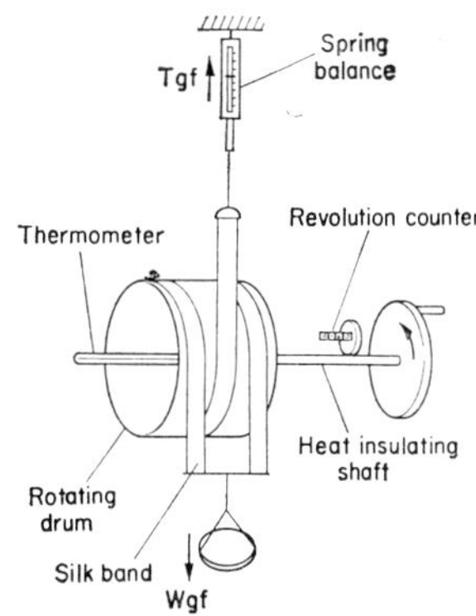

FIG. 30. Mechanical method for J.

S.I. units	*c.g.s. units*
$\dfrac{(W-T)}{1000} gr2\pi n = (Ms + M_1 s_1)\theta$	$\dfrac{(W-T)gr2\pi n}{J} = (Ms + M_1)\theta$
$\therefore\ s_1 = \dfrac{(W-T)gr2\pi n}{M_1 \theta} - \dfrac{Ms}{M_1}$	$\therefore\ J = \dfrac{(W-T)gr2\pi n}{(Ms + M_1)\theta}$

The cooling correction is eliminated by starting with the water a few degrees below room temperature and finishing the same number of degrees above.

An electrical method for the latent heat of vaporization
(Fig. 31)

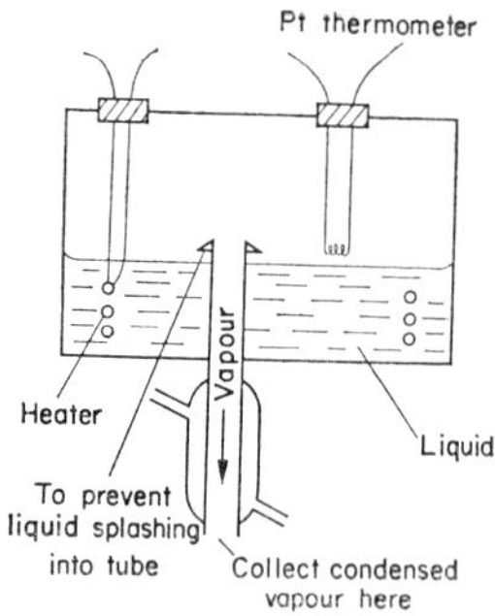

FIG. 31. Electrical method for latent heat of vaporization.

The apparatus is surrounded by an oil bath maintained at the temperature of the boiling liquid. When the conditions are steady all the energy supplied electrically per second (VI watts) is being used to vaporize the liquid. Hence $VI = mL + h$ where m is the number of grams of liquid condensed every second, h is the heat lost or gained from the surroundings, and L is the latent heat in joules per gram. To eliminate h another experiment is conducted when the rate of supply of energy is $V_1 I_1$ watts and the mass of liquid condensed every second is m_1 grams, then $V_1 I_1 = m_1 L + h$. Subtracting the two equations we have

$$V_1 I_1 - VI = (m_1 - m)L$$

and hence L may be determined.

Properties of Gases

Definitions and laws

66. *Boyle's law.* At constant temperature the volume of a fixed mass of gas varies inversely with the pressure.

67. *Charles's law.* At constant pressure the volume of a fixed mass of any gas increases by $\frac{1}{273}$ of its volume at 0°C for each degree C rise in temperature.

68. *Pressure law.* At constant volume the pressure of a fixed mass of any gas increases by $\frac{1}{273}$ of its pressure at 0°C for each degree C rise in temperature.

69. An ideal or perfect gas is one which obeys the gas laws exactly for all temperatures and pressures.

70. The *absolute zero* of temperature is the temperature at which the product pressure times volume for an ideal gas would vanish.

71. The *absolute scale of temperature* (°K) is that scale having its zero at -273°C (the absolute zero) and the size of its degree the same as that on the Celsius scale.

Boyle's law (Fig. 32)

The pressure and volume of the air trapped in the right-hand limb may be changed by raising and lowering the left-hand limb.

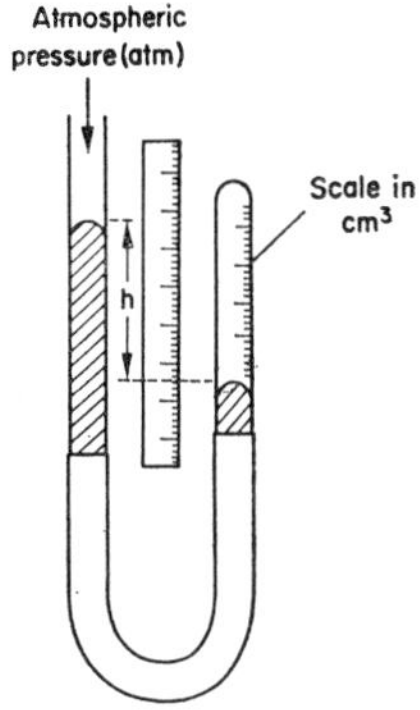

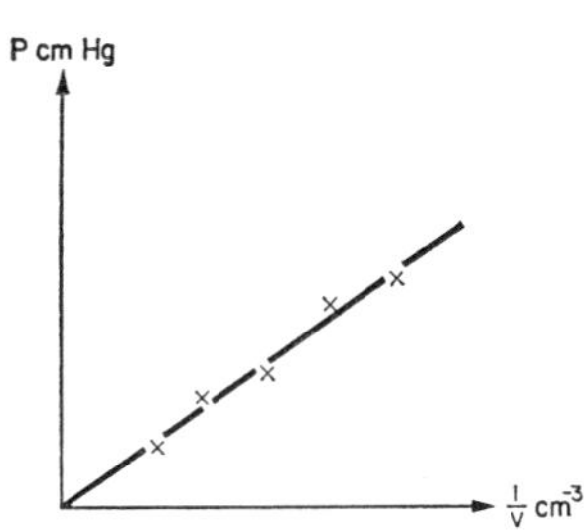

FIG. 32. Boyle's law apparatus.   FIG. 33

Pressure of trapped air $= P = (\text{atm} + h)$ cmHg.

Volume of trapped air $= V$ cubic centimetres.

A graph of P against $1/V$ (Fig. 33) is a straight line through the origin and Boyle's **law** is verified.

Charles's law

Figure 34 is a diagram of Lay's apparatus.

Procedure. (i) Heat water in beaker until temperature rises about 10°C. (ii) Remove bunsen and stir for some minutes. (iii) Level the acid in the central and left-hand limbs. (iv) Record volume and temperature. (v) Obtain a series of readings and plot a graph (Fig. 35) of volume V against temperature t.

If V_t is the volume at $t°C$ and V_0 the volume at $0°C$, then it follows from the definition of coefficient of expansion α that

$$\alpha_P = \frac{V_t - V_0}{V_0 t} = \frac{\text{gradient of graph}}{V_0} \quad \left(= \frac{1}{273} \right)$$

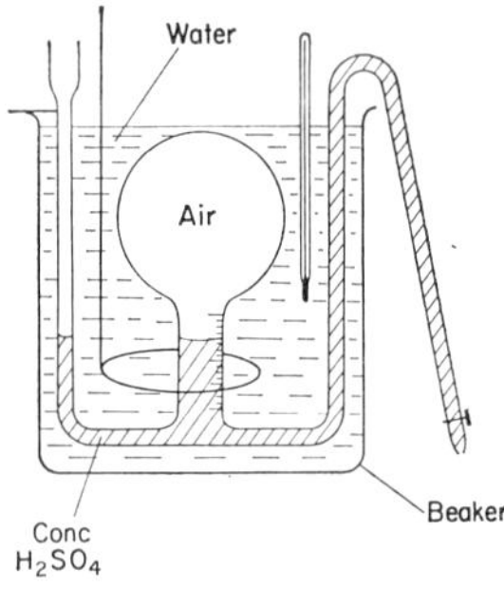

FIG. 34. Lay's apparatus.

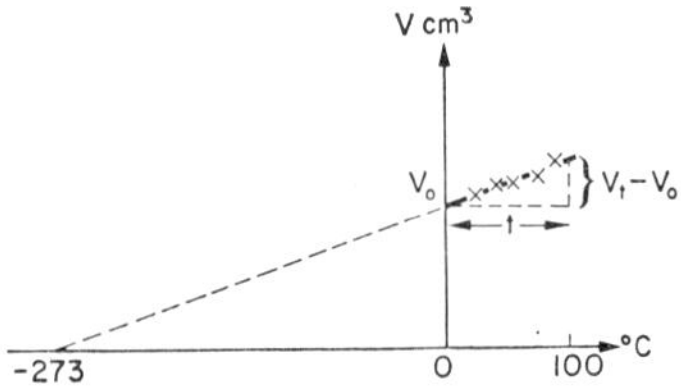

FIG. 35

i.e. the air expands by $\frac{1}{273}$ of its volume at $0°C$ for each degree C rise in temperature. It follows from the above that

$$V_t = V_0 + \frac{t}{273} V_0 = V_0 \left(1 + \frac{t}{273} \right) = V_0 \left(\frac{273 + t}{273} \right) = V_0 \left(\frac{T}{273} \right)$$

where T is the (temperature on the mercury scale) $+ 273°$.

$\therefore V_t/T = V_0/273$, i.e. the volume is proportional to the absolute temperature.

Pressure law

Use the apparatus illustrated in Fig. 36.

Procedure. (i) Heat water in beaker. (ii) Remove bunsen and stir for some minutes. (iii) Bring mercury back to mark M on tube. (iv) Record temperature and the height h.

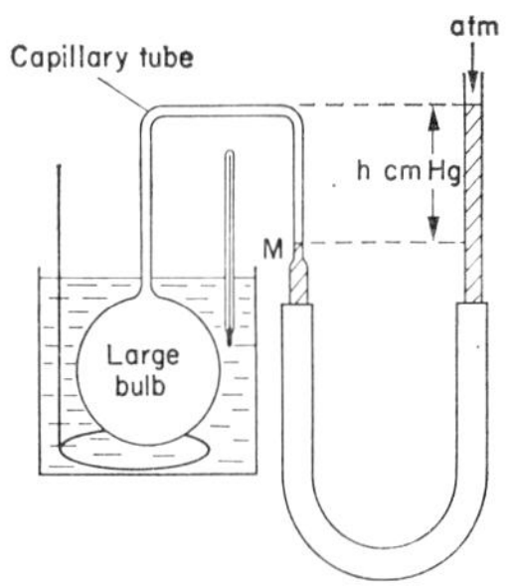

FIG. 36. Experimental verification of pressure law.

Pressure in bulb $= (\text{atm} + h)$ cmHg.

If P_t is the pressure at $t°C$ and P_0 is the pressure at $0°C$ then the pressure coefficient α_v is given by

$$\alpha_v = \frac{P_t - P_0}{P_0 t} = \frac{\text{gradient of graph (Fig. 37)}}{P_0} \quad \left(= \frac{1}{273}\right)$$

The pressure increases by $\frac{1}{273}$ of its pressure at $0°C$ for each degree rise in temperature.

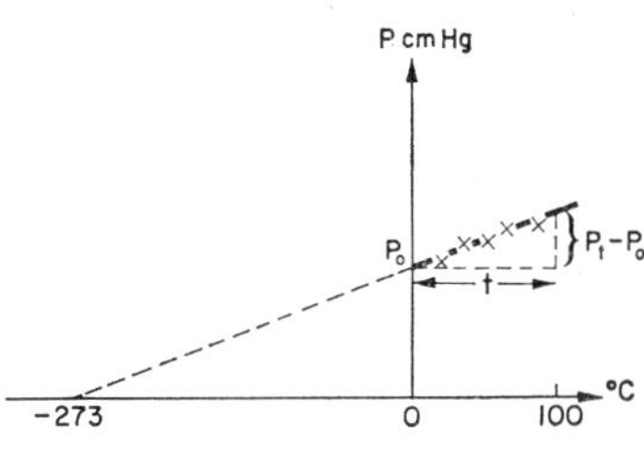

FIG. 37

This is equivalent to the statement that the pressure is proportional to the absolute temperature (for a fixed mass at constant volume). This also follows directly from the fact that the graph is a straight line through the origin if the temperature is measured on the absolute scale.

The perfect gas equation

Consider a fixed mass of gas whose initial pressure volume and absolute temperature are $P_1 V_1 T_1$ and whose final state is represented by $P_2 V_2 T_2$. Let the change take place in two stages (Fig. 38).

(i) A change at constant pressure. If V_1 becomes V and T_1 becomes T_2, then using Charles's law

$$V_1/V = T_1/T_2 \tag{10}$$

(ii) A change at constant temperature. If P_1 becomes P_2 and

V becomes V_2 then using Boyle's law

$$P_1 V = P_2 V_2 \tag{11}$$

Eliminating V from eqns. (10) and (11) we have

$$\frac{P_1 V_1}{T_1} = \frac{P_2 V_2}{T_2}$$

What we have shown is that if we measure temperature on a

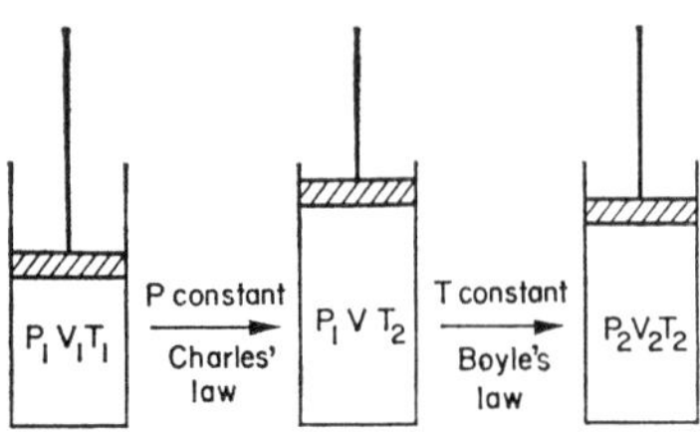

Fig. 38

scale defined by the equation, $T = t°C + 273$, then within the range of the mercury thermometer $P_1 V_1/T_1 = P_2 V_2/T_2$ for air.

An ideal gas is a gas which obeys Boyle's law exactly under all circumstances, and for which the ratio

$$\frac{(pv)_{\text{steam point}}}{(pv)_{\text{ice point}}} = 1\cdot366$$

The variation of pv with temperature is used to define the perfect gas scale of temperature. The perfect gas equation $pv = RT$ is an equation which defines temperature on the perfect gas scale.

Universal gas constant

$PV/T =$ constant, and the value of the constant depends on the mass of gas. If for 1 g of gas $PV/T = r$ then for m grams $PV/T = mr$. For 1 g mol $PV/T = R$ and R is the same constant for all gases, since for 1 g mol V is the same for all gases at S.T.P.

Properties of Vapours

Definitions

72. The *critical temperature* of a substance is the temperature above which the substance cannot be liquefied, however great the pressure.

73. A *gas* is a substance in the gaseous state at a temperature above its critical temperature.

74. A *vapour* is a substance in the gaseous state at a temperature below its critical temperature.

75. An *unsaturated vapour* at any temperature is a vapour not in equilibrium with its liquid at that temperature.

76. A *saturated vapour* at any temperature is a vapour in equilibrium with its liquid at that temperature.

77. The pressure of a saturated vapour at any temperature is called the *saturated vapour pressure* of the substance at that temperature.

78. A liquid boils when its vapour pressure is equal to the external pressure.

79. *Dalton's law of partial pressures.* In a mixture of gases each gas exerts a pressure which is the same as if it were alone occupying the volume of the mixture.

Evaporation and boiling

Evaporation is a surface phenomenon, the faster moving molecules escaping from the surface. Boiling takes place in the body of the liquid when the vapour pressure is equal to the external pressure (the J-tube experiment demonstrates this).

Saturated vapours

If a liquid is in contact with its vapour in a small closed space the vapour will be saturated. A state of dynamic equilibrium exists, that is, the rate at which molecules are leaving the surface is equal to the rate at which molecules are returning to the surface.

Measurement of S.V.P.

(i) *Static method* 0–60°C *for water* (Fig. 39). A small quantity of water is introduced into the left-hand barometer tube using a bent pipette. The height h is the S.V.P. In accurate work allowance must be made for (a) the weight of water in the tube, (b) the surface tension depression is not the same in each limb.

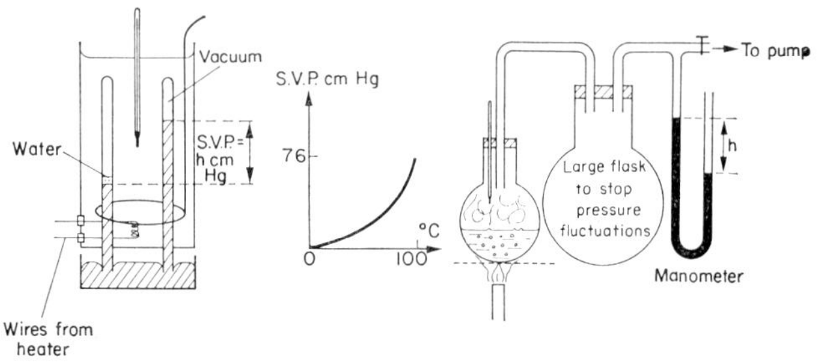

Fig. 39. Measurement of S.V.P.   Fig. 40. Measurement of S.V.P.

(ii) *Dynamic method* (Regnault used it from 0 to 200°C for water). When the water in the flask (Fig. 40) boils, the S.V.P. = pressure above boiling water = $(\mathrm{atm} - h)$ cmHg. *Procedure.* Alter pressure above water by using the pump. Read the thermometer when the liquid is boiling and the manometer height h centimetres.

Dalton's law of partial pressures

In practice this law means that we can apply $PV/T = \mathrm{constant}$, to the air in a mixture of air and water vapour.

Water vapour in the atmosphere. Definitions

80. The *dew point* is the temperature at which the atmosphere, at any particular moment, would be just saturated with water vapour.

81. The *absolute humidity* is the mass of water vapour in a cubic centimetre of air.

82. The *relative humidity* of the atmosphere is the ratio of the mass of water vapour in a given volume, to the mass of water vapour required to saturate the same volume at the same temperature.

Determination of relative humidity from dew point

It may be shown that the relative humidity is equal to the ratio of the saturated vapour pressure at the dew point, to the saturated vapour pressure at the air temperature. The vapour pressure is found from tables, and the dew point by observing the temperature when dew forms on the metal cap of a test-tube when air is bubbled through some ether in the tube. Observation is best made through a telescope.

Wet and dry bulb hygrometer

Two thermometers are arranged side by side, and the bulb of one of them is covered with a piece of moist muslin. Tables are available which give the relative humidity in terms of the difference of the thermometer readings.

Kinetic Theory of Gases

Consider a cube of side 1 cm containing n molecules each of mass m. The velocity of a typical particle c is the vector sum of the components u, v and w (Fig. 41).
$c^2 = u^2 + v^2 + w^2$.

Momentum change of this molecule on rebounding from face B (assuming that the molecules are perfectly elastic and that the range of intermolecular forces

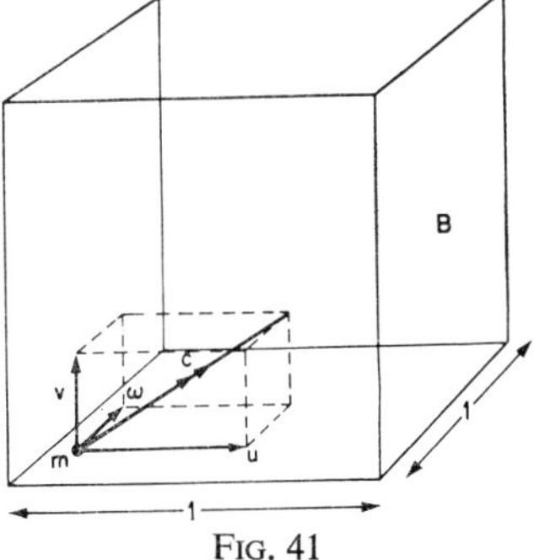

FIG. 41

is small compared with the mean distance between the molecules)
$= 2mu$.

Number of collisions with B per second (assuming the size of the molecules and the collision time are negligible) $= u/2$.

Change of momentum per second for this molecule

$$= 2mu \, . \, u/2 = mu^2.$$

Let the average value of u^2 for all the n molecules be $\overline{u^2}$. Then pressure on B (by Newton's second law) $= nm\overline{u^2}$.

But experiment shows that the pressure on every wall of the container is the same. Hence $nm\overline{u^2} = nm\overline{v^2} = nm\overline{w^2}$ and therefore $\overline{u^2} = \overline{v^2} = \overline{w^2}$. If $\overline{c^2}$ is the average of the squares of all the velocities, then $\overline{c^2} = \overline{u^2} + \overline{v^2} + \overline{w^2} = 3\overline{u^2}$.

Pressure (P) on $B = nm\overline{u^2} = \frac{1}{3}nm\overline{c^2} = \frac{1}{3}\rho\overline{c^2}$ (since $nm = \rho =$ density of gas). For a quantity of gas of mass M occupying a volume V we have $P = (\frac{1}{3}M/V)\overline{c^2}$ or $PV = \frac{1}{3}M\overline{c^2}$. If $\overline{c^2}$ depends only on T, the absolute temperature, then

$$PV = \text{constant}$$

at constant temperature, i.e. Boyle's law.

Further, if $\overline{c^2} \propto T$ then $PV = RT$ (where R is a constant), i.e. the gas equation.

Brownian movement

If a suspension of rubber latex solution in water is viewed through a microscope the particles in suspension are in incessant random motion. This motion is evidence for the kinetic theory of liquids, since in any small time interval a particle will receive more impacts from the liquid molecules on one side than another. Millikan's apparatus (page 113) provides visible evidence of Brownian movement for an oil drop in air.

Experiment to demonstrate the rapid velocity of
bromine molecules

If a phial of bromine is broken inside an evacuated gas jar, the colour spreads almost instantaneously throughout the gas jar. If the jar is not evacuated, the bromine vapour takes some minutes to diffuse.

Thermometry

Definitions

83. The temperature in degrees Celsius is defined by the equation

$$t°\text{C} = \frac{\text{change in magnitude of the property}}{\frac{1}{100}\ \text{(change in magnitude in going from the lower fixed point to the upper fixed point)}}$$

84. The fundamental temperature scale is arbitrarily chosen as the ideal gas scale. The equation $pv = RT$ defines a temperature T on this scale.

Two scales of temperature will only agree if the property on which one is based varies uniformly with temperature measured according to the other.

Constant volume hydrogen thermometer

Figure 36 is a constant volume gas thermometer. From the definition of temperature in degrees Celsius

$$t°\text{C} = \frac{p - p_0}{\frac{1}{100}(p_{100} - p_0)}$$

where p, p_{100} and p_0 are the pressures when the gas in the bulb is at the unknown temperature, at the upper fixed point and at the ice point respectively. In each case the pressure is found from the relationship $p = (\text{atm} + h)$ cmHg. Allowance must be made for (i) the volume of the dead space, and (ii) the change in volume of the bulb as the temperature changes.

Platinum resistance thermometer (Fig. 42)

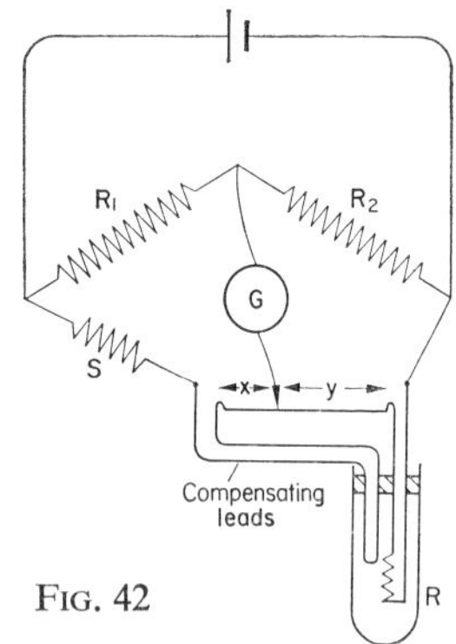

FIG. 42

The compensating leads are to ensure that the change in resistance of the leads to the thermometer does not affect the balance point. S is a resistor approximately equal to R. Applying eqn. (17) on page 77

$$\frac{R_1}{R_2} = \frac{S + R_L + x\rho}{R + R_L + y\rho}$$

where R_L is the resistance of the leads and ρ is the resistance of unit length of bridge wire. In practice R_1 is made equal to R_2 and the above equation then gives

$$R = S + (x - y)\rho$$

The temperature is found by substituting in the equation

$$t°C = \frac{R - R_0}{\frac{1}{100}(R_{100} - R_0)}$$

Callendar showed that for $t°C$ measured on the constant volume hydrogen scale $R_t = R_0(1 + at + bt^2)$ where a and b are constants. Temperatures may be converted to the constant volume hydrogen scale using this equation.

Small beads of special semiconductors, known as thermistors, are now often used as resistance thermometer elements. Their resistance *decreases* rapidly with a rise in temperature.

Thermocouple thermometer

The hot junction acts as the temperature probe, the cold junction being kept at a constant temperature. The thermoelectric e.m.f. is measured using the potentiometer circuit illustrated in Fig. 93, page 80. The international temperature corresponding

Type	Advantages	Disadvantages	Range
Liquid in glass	(1) Simple to use. (2) Direct reading. (3) Easy to read. (4) Convenient.	(1) Small range. (2) Zero changes with time. (3) Reading depends on length of exposed column. (4) Liquid sticks to side of tube. (5) Variation of pressure on bulb alters reading.	$-40°C$ to $500°C$ with nitrogen above mercury. Other ranges using different liquids.
Gas thermometer	(1) Very large range. (2) Accurate because of large gas expansion. (3) Agrees with the Kelvin thermodynamic scale if extrapolated to ideal behaviour $(p \to 0)$.	(1) Bulky. (2) Slow in action. (3) Uncertain correction for dead space. (4) Not direct reading. (5) Allowance must be made for expansion of bulb.	$-250°C$ to $2000°C$.
Platinum resistance thermometer	(1) Reads easily to $\frac{1}{50}$ °C. (2) Good range. (3) Measures temperature differences accurately when connected differentially in pairs.	(1) Not direct reading. (2) Rather complicated.	$-200°C$ to $1100°C$.
Thermocouple	(1) Very low thermal capacity and small size ensures minimum disturbance. (2) Good range. (3) Quick in action.	(1) Not direct reading. (2) E.m.f. affected by impurities in wire.	$-250°C$ to $1800°C$.

to an e.m.f. E is given by $E = a + bt + ct^2$. The values of the constants a, b and c are found by measurements at a number of internationally agreed fixed points.

Thermal Conductivity

Definition

85. The rate of conduction of heat $(\delta Q/\delta t)$ normally through a surface of area A is given by

$$\frac{\delta Q}{\delta t} = -kA\frac{\delta\theta}{\delta x}$$

where $\delta\theta/\delta x$ is the temperature gradient and k is the *thermal conductivity*. At ordinary temperatures k is independent of A and $\delta\theta/\delta x$ but it varies with θ. A mean value of k may be used provided the range of θ is not too great (usually watts $\text{degC}^{-1}\text{m}^{-1}$ or $\text{cal s}^{-1}\,\text{degC}^{-1}\text{cm}^{-1}$).

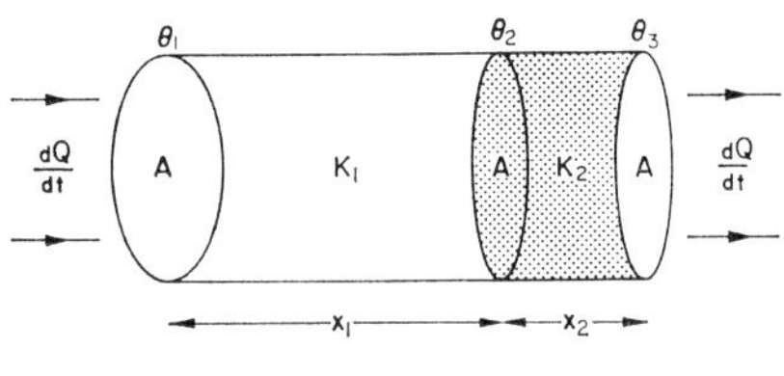

FIG. 43

Conduction through composite walls

In Fig. 43 x_1 and x_2 are the thicknesses of the materials having thermal conductivities k_1 and k_2. θ_1, θ_2 and θ_3 are the temperatures at the planes indicated in the diagram. Provided no heat is lost from the sides, the rate of flow of heat (dQ/dt) across each area A is the same and is given by

$$\frac{dQ}{dt} = k_1 A\frac{\theta_1-\theta_2}{x_1} = k_2 A\frac{\theta_2-\theta_3}{x_2}$$

Measurement of thermal conductivity

(a) *Good conductors.* *Searle's bar* (Fig. 44). When the conditions are *steady* then the heat flowing along the bar per second is equal to the heat taken up by the cold water flowing through the pipes at the far end:

$$\frac{dQ}{dt} = kA\,\frac{\theta_1-\theta_2}{x} = ms\,(\theta_4-\theta_3)$$

where m is the mass of water flowing per second, s the specific heat of water, k the thermal conductivity of the bar, A the cross-sectional area, and θ_1, θ_2, θ_3 and θ_4 the temperatures shown in Fig. 44.

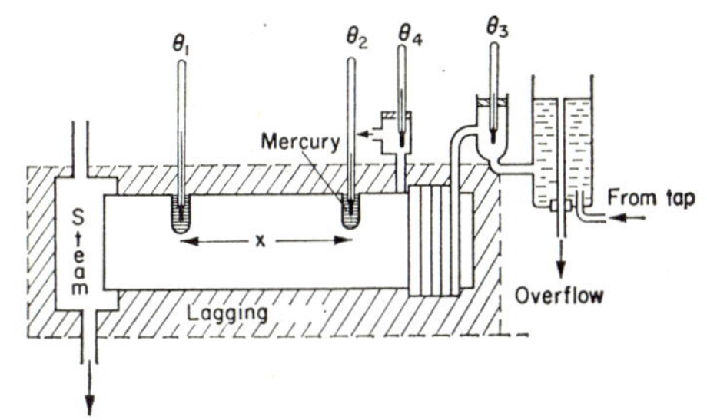

FIG. 44. Searle's apparatus.

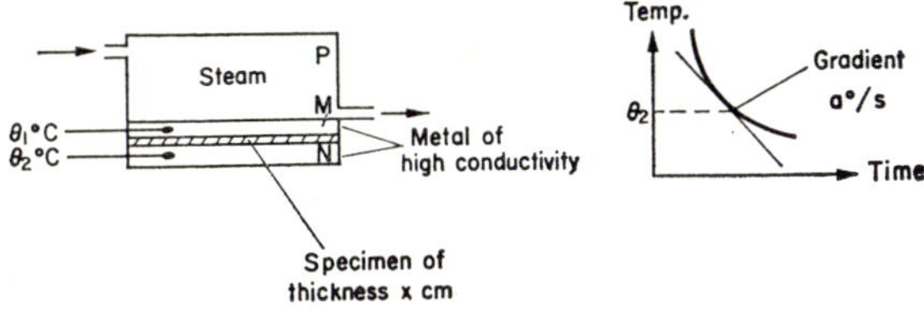

FIG. 45. Lees's disc.

(b) *Poor conductors.* *Lees's disc* (Fig. 45). The specimen in the form of a thin disc is sandwiched between two good conducting discs. These discs serve to equalize the temperature over the surface of the specimen, and thermometers are inserted to measure the temperature difference. Neglecting the heat losses from the sides of the specimen, which are very small, the rate of flow of heat per second across it is given by

$$\frac{dQ}{dt} = kA\frac{(\theta_1 - \theta_2)}{x}$$

The rate of flow of heat is also the rate of loss of heat from N to the surrounding air. This may be determined by removing P and M and coating the top of the specimen with a varnish of low emissivity. Heat N to a temperature above θ_2 and allow the apparatus to cool. Plot a cooling curve, and the gradient of the tangent at the temperature θ_2 gives the rate of fall of temperature when N is at a temperature θ_2. If this is a degrees per second, then the rate of loss of heat from N is msa, where m is the mass of N and s its specific heat. k may be calculated from the equation

$$kA\frac{\theta_1 - \theta_2}{x} = msa$$

Specific Heats of Gases

Definitions

86. The *specific heat of a gas at constant volume* c_v is the quantity of heat required to raise the temperature of unit mass of gas through 1 degree, the volume being kept constant (usually $J\ kg^{-1}\ degC^{-1}$ or $cal\ g^{-1}\ degC^{-1}$).

87. The *specific heat of a gas at constant pressure* c_p is the quantity of heat required to raise the temperature of unit mass of gas through 1 degree, the pressure being kept constant throughout (usually $J\ kg^{-1}\ degC^{-1}$ or $cal\ g^{-1}\ degC^{-1}$).

Work done when a gas expands

Suppose in Fig. 46 the gas at a pressure p expands and pushes the piston back a distance δx. If the area of the piston is A, then the work done is $pA \times \delta x = p\,.\,\delta v$, i.e. p times the change in volume.

Difference between the specific heats of a gas

Consider 1 g of an ideal gas heated through $\delta T°C$ at constant pressure. The heat energy supplied is $c_p\delta T$. This is the energy to raise the temperature of the gas $\delta T°C$ ($c_v\delta T$) and the energy to do the work involved in expansion ($p\delta v$). Therefore

$$c_p\delta T = c_v\delta T + p\delta v$$

Since $pv = rT$, then at constant pressure $p\delta v = r\delta T$. Substituting in the previous equation we have

$$c_p\delta T - c_v\delta T = r\delta T \quad \text{or} \quad c_p - c_v = r$$

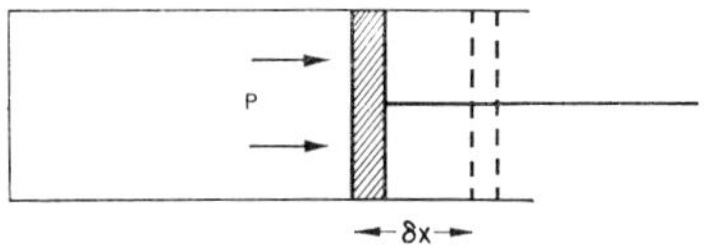

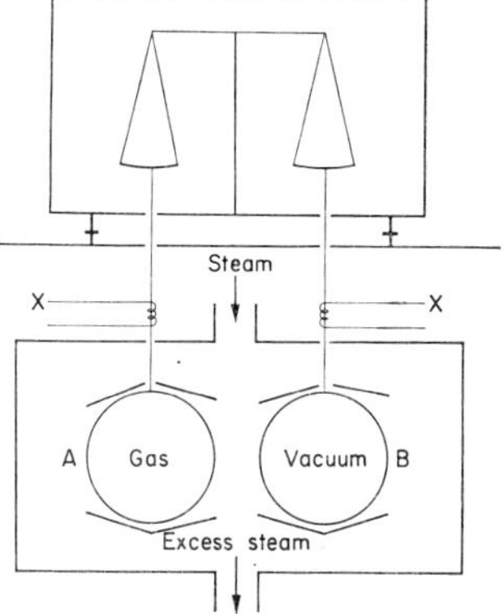

FIG. 46

Determination of c_v. Joly's steam calorimeter (Fig. 47)

If W grams of gas are in sphere A and an excess weight of water w condenses on that sphere, then $wL = Wc_v(100-\theta)$ where θ is the initial temperature of the gas and L the latent heat of vaporization of water. The coils XX prevent water drops forming on the wire. In accurate work allowance must be made for (i) the unequal thermal capacities of the spheres. A second experiment is conducted with sphere B full of gas and sphere A

FIG. 47. Joly's steam calorimeter.

evacuated. (ii) The work done by the gas in expanding (resulting from the thermal expansion of the sphere). The temperatures are measured by a thermometer in the chamber. This method is now only of historical interest. A modern method is described on page 133.

Determination of c_p

Any heat escaping from the heater (Fig. 48) warms the incoming gas. The wire gauze is to ensure the gas is at uniform temperature before it passes the platinum resistance thermometer. Let m be the mass of gas flowing per second and the temperature rise $(\theta_2 - \theta_1)$. The rate of supply of heat is $VI = mc_p(\theta_2 - \theta_1) + h$. As in Callendar and Barnes's experiment for a liquid the heat loss h may be eliminated by conducting a second experiment with a different rate of supply of heat. A valve connected to the supply cylinder is adjusted to keep the pressure constant. A knowledge of the initial and final pressure of the gas in the cylinder enables the initial and final density and hence the mass of gas passed to be calculated.

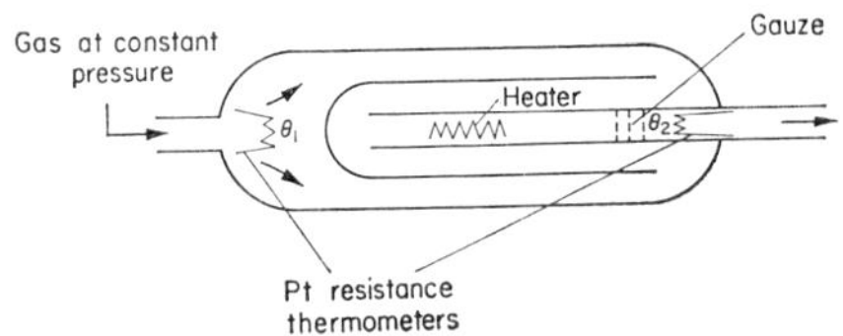

FIG. 48. Constant flow method for c_p.

Isothermal and Adiabatic Expansion

Definitions

88. An *isothermal* change is a change at constant temperature.

89. An *adiabatic* change is a change during which no heat enters or leaves the system.

Laws which apply to these changes

For an isothermal change $pv = $ constant, but for an adiabatic reversible change $pv^\gamma = $ constant. The work done during a change may be found by integrating $p\delta v$. $pv = rT$ applies both to an adiabatic and an isothermal change. The above equations are only strictly true for an ideal gas.

Deviations from the Gas Laws

Boyle's law is obeyed only by an ideal gas. The curves in Fig. 49 indicate what happens for an actual gas. Only at a certain temperature, the Boyle temperature, is *pv* a constant over any reasonable range of pressure.

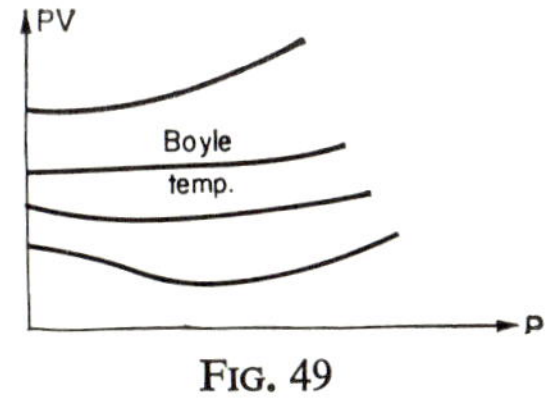

FIG. 49

Andrews's experiments on carbon dioxide (Fig. 50)

Procedure. (i) Pass CO_2 through tube for about 24 hr and seal one end. (ii) Put under mercury and reduce pressure of air above the mercury. (iii) Screw up plungers at bottom, this forces the

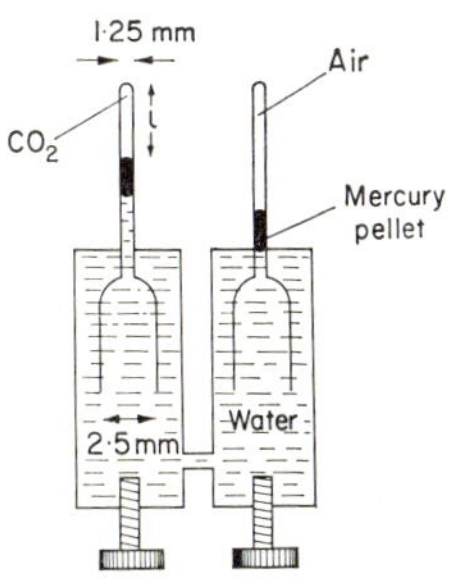

FIG. 50. Andrews's apparatus.

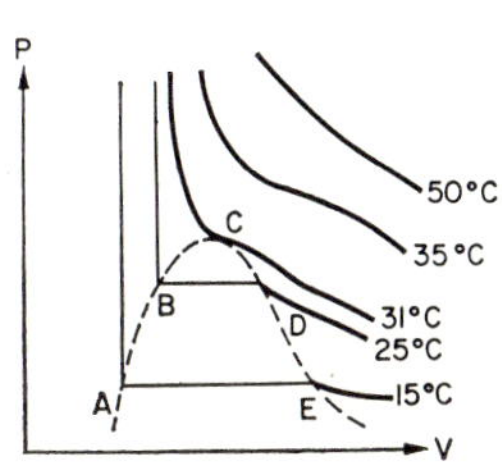

FIG. 51

mercury thread into the capillary tube. (iv) Boyle's law is applied to the air in the adjacent tube in order to determine the pressure of the enclosed CO_2. The volume of the CO_2 is proportional to the length *l*. *Results.* These are shown in Fig. 51. Above about 50°C Boyle's law is approximately obeyed. In the state represented by the inside of the dotted curve *ABCDE* there is a mixture of liquid and vapour.

Van der Waals's equation

When deriving the equation $PV = RT$ using the kinetic theory, it was assumed that the range of intermolecular forces was small compared with the separation of the molecules and that the volume occupied by the molecules was negligible. Both these assumptions are invalid as the pressure increases. In the equation

$$\left(p+\frac{a}{v^2}\right)(v-b) = RT$$

where a and b are constants, the term a/v^2 allows for the attractive forces between the molecules, and b the short range repulsive forces (i.e. the volume in which the molecules are free to move is less than the volume of the container).

CHAPTER 3

Light

Reflection and Refraction

Definitions and laws

90. *Laws of reflection.* (a) The reflected ray lies in the plane formed by the incident ray and the normal to the reflecting surface at the point of incidence. It lies on the opposite side of the normal to the incident ray.

(b) The angle between the reflected ray and the normal equals the angle between the incident ray and the normal.

91. *Laws of refraction.* (a) The refracted ray lies in the plane formed by the incident ray and the normal to the refracting surface at the point of incidence. It lies on the opposite side of the normal to the incident ray.

(b) *Snell's law.* When a ray passes from medium A to medium B the ratio of the sine of the angle of incidence to the sine of the angle of refraction for light of a particular wavelength is constant and is termed the refractive index from A to B. Mathematically $_1\mu_2 = \sin i/\sin r$ or if μ is the absolute refractive index, $\mu \sin i = $ constant.

92. The *absolute refractive index* of a medium is the ratio of the sine of the angle of incidence to the sine of the angle of refraction when light passes from a vacuum to the medium.

93. The *critical angle* for any two media is the angle of incidence of light on the boundary (in the optically denser medium) such that the angle of refraction is $90°$.

Relationships between refractive indices

$$_1\mu_2 = \frac{1}{_2\mu_1} \quad \text{and} \quad _1\mu_3 = {_1\mu_2} \cdot {_2\mu_3}$$

Real and apparent depth (Fig. 52)

An object O gives rise to an image I. Applying Snell's law

$$_1\mu_2 = \frac{\sin x}{\sin y} = \frac{AB/AI}{AB/AO}$$

$$= \frac{AO}{AI} \simeq \frac{OB}{BI} \text{ if viewed normally,}$$

i.e. μ (from where the eye is to where you are looking) $= \dfrac{\text{real depth}}{\text{apparent depth}}$

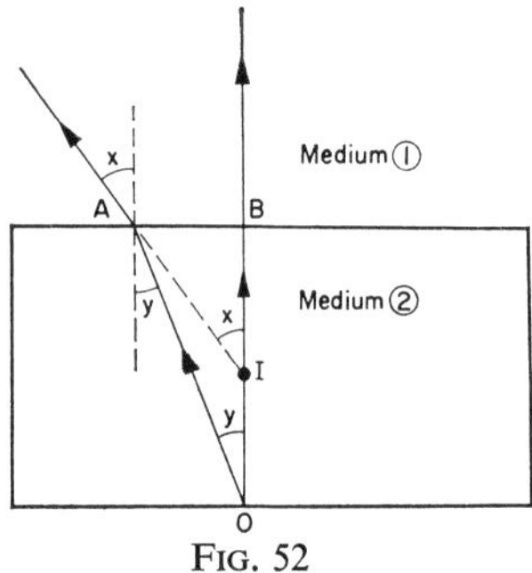

FIG. 52

Measurement of refractive index of liquids

(i) *Travelling microscope method.* Focus the microscope on the bottom of the container. Put some liquid in the container and refocus the microscope on the bottom. Focus the microscope on the surface of the liquid. If the readings are a, b and c respectively then

$$\mu = \frac{\text{real depth}}{\text{apparent depth}} = \frac{c-a}{c-b}$$

The method is also suitable for the refractive index of the material of a glass or plastic block.

(ii) *Air cell method.* Two thin glass plates are cemented together so as to contain an air film of constant thickness between them. The system is immersed in a liquid of refractive index μ_l and rotated until light on one face just ceases to be transmitted.

Then using the notation in Fig. 53:

$$\mu_l \sin i_1 = \mu_g \sin i_2 = \mu_a \sin 90$$

$$\mu_l/\mu_a = 1/\sin i_1$$

$$_a\mu_l = 1/\sin i_1$$

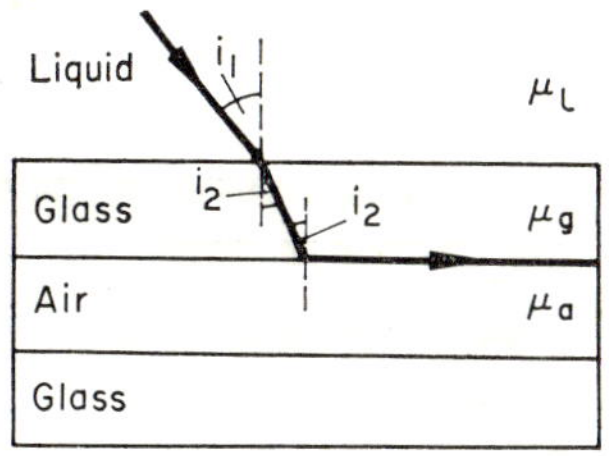

FIG. 53

In practice the angle $2i_1$ is measured because this is more accurate.

Prisms

At minimum deviation (the symmetrical ray) Fig. 54

$$A = r+r = 2r$$

and

$$D = (i-r)+(i-r) = 2i-A \quad \therefore i = \tfrac{1}{2}(A+D)$$

But

$$\mu = \frac{\sin i}{\sin r} = \frac{\sin \tfrac{1}{2}(A+D)}{\sin \tfrac{1}{2}A} \tag{12}$$

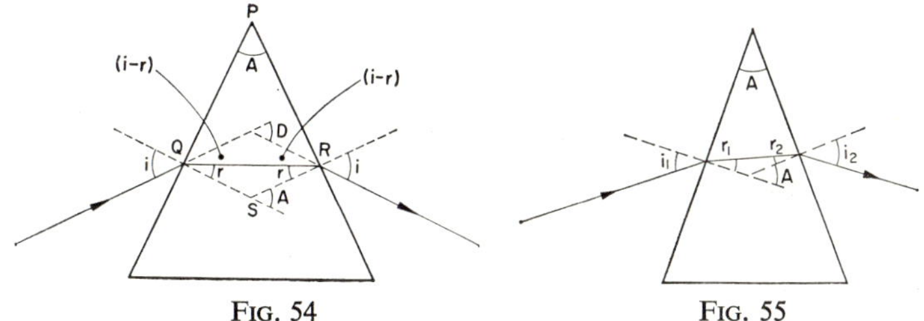

FIG. 54   FIG. 55

For prisms of small angle and for small angles of incidence we have (Fig. 55) $i_1 = \mu r_1$ and $i_2 = \mu r_2$. Substituting in

$$d = (i_1-r_1)+(i_2-r_2)$$

$$d = (\mu r_1-r_1)+(\mu r_2-r_2) = \mu(r_1+r_2)-(r_1+r_2)$$

But $r_1+r_2 = A$.

$$\therefore \quad d = (\mu-1)A$$

Curved Mirrors and Lenses

Definitions. Spherical mirrors

94. The *centre of curvature* is the centre of the sphere of which the mirror is part.

95. The *aperture* is the length of the chord drawn from the extremities of the mirror.

96. The *pole* of the mirror is the midpoint of the mirror.

97. The *principal axis* is the line through the centre of curvature and the pole.

98. Rays parallel and near to the principal axis, after reflection at the mirror, converge to or appear to diverge from a point on the axis, known as the *principal focus.*

99. The *focal length* is the distance from the pole to the principal focus.

Thin lenses

100. The *centre of curvature* of a lens surface is the centre of the sphere of which the surface is part.

101. The *principal axis* is the line joining the centres of curvature.

102. Rays parallel and near to the principal axis, after refraction, converge to or appear to diverge from a point on that axis known as the *principal focus.*

103. A ray of light which passes through the lens undeviated cuts the principal axis at a point called the *optical centre.*

104. The *focal length* is the distance from the optical centre to a principal focus.

105. A *real object, image or principal focus* is one through which light actually passes: a *virtual object, image or principal focus* is one through which light only appears to pass.

106. *Convention of signs.*

Real is positive (R.P.)	*New Cartesian* (N.C.)
The distance from a lens (or mirror) to a real object, image or principal focus is positive, and the distance from a virtual object, image or principal focus is negative. For mirrors the radius of curvature has the same sign as the focal length. For lenses the radius of curvature is positive if the surface is convex towards the less dense medium.	All distances are measured from the lens or mirror as origin. Distances measured against the incident light are negative and distances measured in the same direction as the incident light are positive. Therefore if the object is on the left the ordinary graphical convention of signs applies, i.e. distances to the left are negative, those to the right are positive.

107. The *power* of a lens is the reciprocal of the focal length.

108. The power of a lens is one *dioptre* if the focal length is one metre.

Relation between focal length and radius of curvature

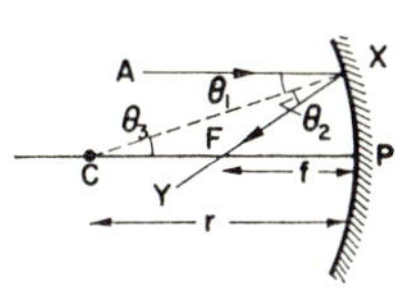

FIG. 56

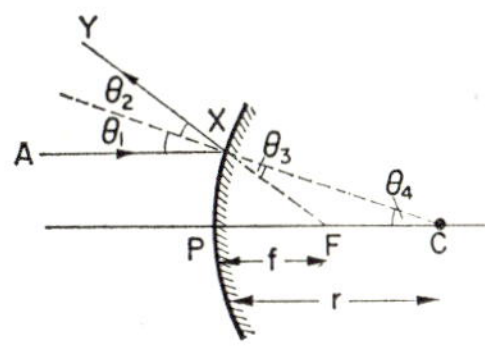

FIG. 57

Consider a ray AX parallel to the principal axis. If CX is the normal to the mirror at X and XY is the reflected ray then:

(a) *for a concave mirror*

$\theta_1 = \theta_2$ and $\theta_1 = \theta_3$

$\therefore \theta_2 = \theta_3$

(b) *for a convex mirror*

$\theta_1 = \theta_2$ and $\theta_1 = \theta_4$ and $\theta_2 = \theta_3$

$\therefore \theta_3 = \theta_4$

Therefore $XF = FC$. For paraxial rays $XF = PF$ to a very good approximation.

$$\therefore PF = FC \quad \text{i.e.} \quad f = r/2$$

Relationship between u, v and f

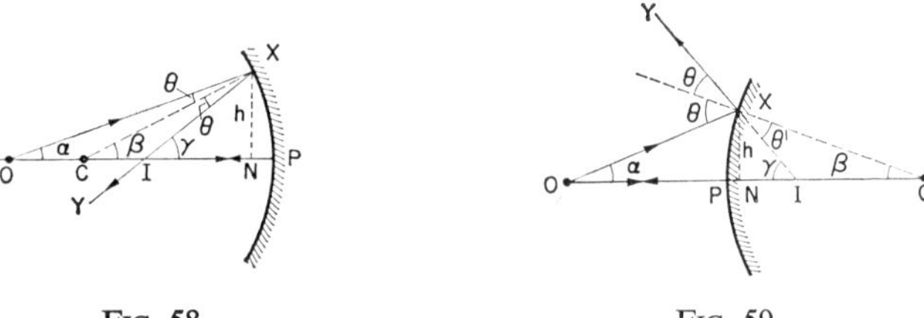

FIG. 58   FIG. 59

Consider rays OX and OP from a point object at O. Let the reflected rays be XY and PO. Since these intersect at I, I is the image of O. The angles marked θ are equal (laws of reflection).

Concave mirror	*Convex mirror*
$\gamma = \theta + \beta, \quad \beta = \theta + \alpha$	$\theta' = \theta, \quad \gamma = \theta' + \beta, \quad \theta = \beta + \alpha$
$\therefore \gamma = 2\beta - \alpha \quad \dots \text{(i)}$	$\therefore \gamma = 2\beta + \alpha \quad \dots \text{(ii)}$

If N is the foot of the perpendicular from X on to the principal axis, then for paraxial rays $\gamma = h/IN$, $\beta = h/CN$, $\alpha = h/ON$ and since N and P approximately coincide $\gamma = h/IP$, $\beta = h/CP$, $\alpha = h/OP$. Substituting in eqns. (i) and (ii)

$$h/IP = 2h/CP - h/OP \qquad h/IP = 2h/CP + h/OP$$

Applying the sign convention stated on page 53:

R.P.	N.C.	R.P.	N.C.
$u = +OP$	$u = -OP$	$u = +OP$	$u = -OP$
$v = +IP$	$v = -IP$	$v = -IP$	$v = +IP$
$r = +CP$	$r = -CP$	$r = -CP$	$r = +CP$

$$1/u + 1/v = 1/f = 2/r \tag{13}$$

Linear magnification

It follows from the geometry of Fig. 60 that

$$\text{linear magnification } m = h_2/h_1 = |v/u|$$

Substituting in eqn. (13)

$$m = \frac{v}{f} - 1 \quad \text{and} \quad \frac{1}{m} = \frac{u}{f} - 1$$

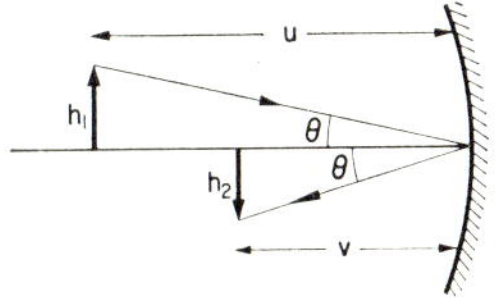

FIG. 60

Thin lenses

A thin lens is effectively a number of small angled prisms of gradually changing angle. In Fig. 61 C_1 and C_2 are centres of

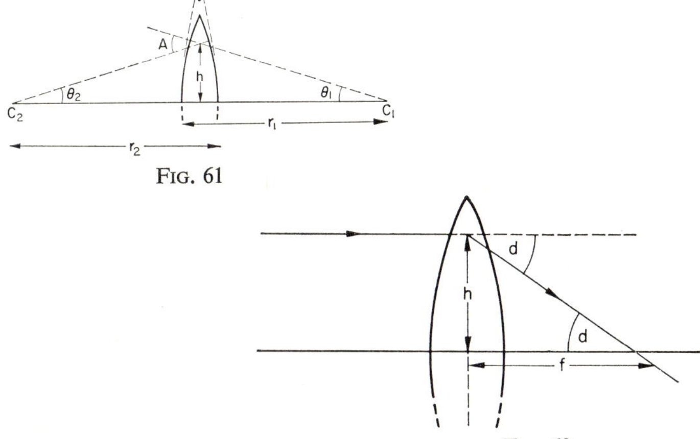

FIG. 61

FIG. 62

curvature of the lens faces. The radii of curvature shown in the diagram intersect at an angle A equal to the angle of the prism.

R.P.	N.C.

$A = \theta_1 + \theta_2 = h/r_1 + h/r_2$

But

$$d = (\mu - 1)A$$
$$= (\mu - 1)(h/r_1 + h/r_2)$$

Reference to Fig. 62 shows that

$$d = h/f.$$

Therefore

$$h/f = (\mu - 1)(h/r_1 + h/r_2)$$

and hence for all values of h

$$\underline{1/f = (\mu - 1)(1/r_1 + 1/r_2)}$$

Referring to Fig. 63

$$\delta = \alpha + \beta = h/u + h/v$$

But $\delta = h/f$ (see Fig. 62)

$$\therefore \underline{\frac{1}{f} = \frac{1}{u} + \frac{1}{v}}$$

$A = \theta_1 + \theta_2 = h/r_1 - h/r_2$

But

$$d = (\mu - 1)A$$
$$= (\mu - 1)(h/r_1 - h/r_2)$$

Reference to Fig. 62 shows that

$$d = h/f.$$

Therefore

$$h/f = (\mu - 1)(h/r_1 - h/r_2)$$

and hence for all values of h

$$\underline{1/f = (\mu - 1)(1/r_1 - 1/r_2)}$$

Referring to Fig. 63

$$\delta = \alpha + \beta = (h/-u) + (h/+v)$$

But $\delta = h/f$ (see Fig. 62)

$$\therefore \underline{\frac{1}{f} = \frac{1}{v} - \frac{1}{u}} \tag{14}$$

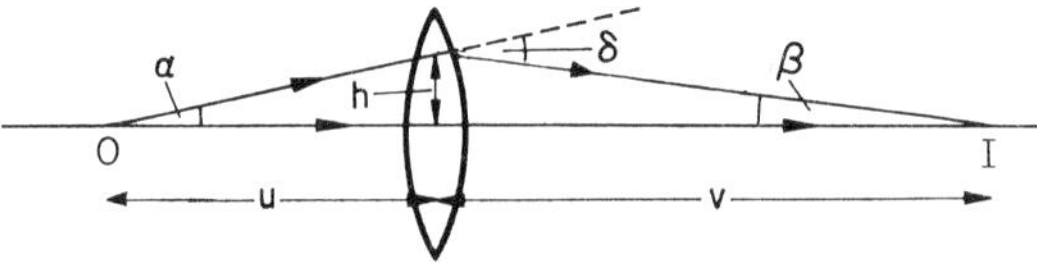

FIG. 63

Experimental determination of r (and f) for a mirror

Concave mirror. (i) Coincident object and image. Both are at a distance r from the mirror. (ii) Measure u and v for a series of object and image distances. Plot $1/u$ against $1/v$ and the intercepts are $1/f$, or, plot m against v and the gradient is $1/f$.

Convex mirror. (i) Locate the virtual image using a plane mirror. u and v are then known and substitution in eqn. (14) gives f.

(ii) Use a converging lens and adjust the position of the mirror until there is a coincident object and image (Fig. 64).

Experimental determination of f for a lens

Converging. (i) Focus a distant object on to a screen. (ii) Place a plane mirror behind the lens. When a coincident object and image are formed, both are at the focus of the lens. (iii) Measure u and v and plot graphs as for mirrors.

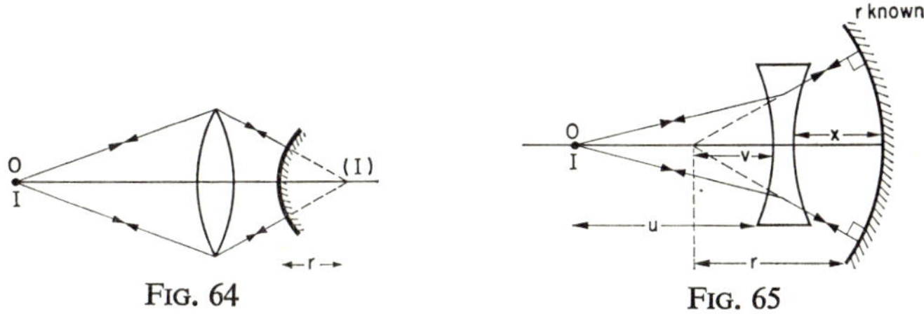

FIG. 64   FIG. 65

Diverging. (i) Locate the virtual image using a plane mirror. (ii) Use a concave mirror of known radius of curvature and produce a coincident object and image as illustrated in Fig. 65. Since the image distance is $(r-x)$, u and v are known and substitution in eqn. (14) gives f.

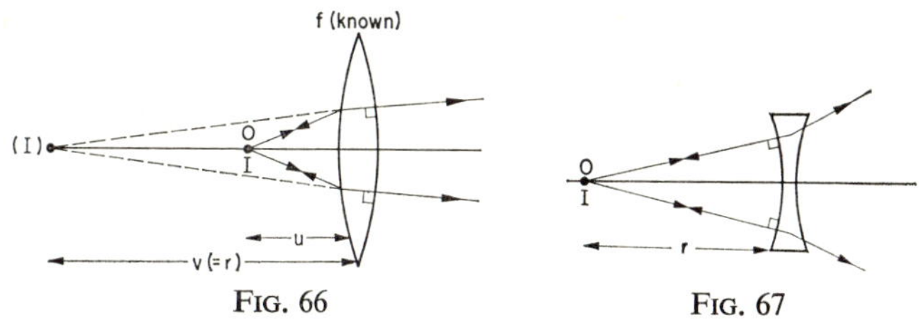

FIG. 66   FIG. 67

Experimental determination of r for a lens

Converging lens. Boys's method (Fig. 66). Locate the coincident object and image, the image being formed by light reflected from the second surface of the lens. Knowing u and f, v is calculated, and this is equal to r.

Diverging. Use the surface of the lens as a mirror (Fig. 67).

Two lenses in contact

If u is the object distance, v the image distance and f_1 and f_2 the focal lengths of the two lenses, then for the first lens

<table>
<tr><td align="center">R.P.</td><td align="center">N.C.</td></tr>
</table>

$$\frac{1}{u} + \frac{1}{v_1} = \frac{1}{f_1} \quad \text{(i)} \qquad\qquad \frac{1}{v_1} - \frac{1}{u} = \frac{1}{f_1} \quad \text{(i)}$$

where v_1 is the image distance in the absence of the second lens. For the second lens v_1 is the object distance

$$\frac{1}{-v_1} + \frac{1}{v} = \frac{1}{f_2} \quad \text{(ii)} \qquad\qquad \frac{1}{v} - \frac{1}{v_1} = \frac{1}{f_2} \quad \text{(ii)}$$

adding (i) and (ii)   adding (i) and (ii)

$$\frac{1}{u} + \frac{1}{v} = \frac{1}{f_1} + \frac{1}{f_2} \qquad\qquad \frac{1}{v} - \frac{1}{u} = \frac{1}{f_1} + \frac{1}{f_2}$$

$$\therefore \ 1/F = 1/f_1 + 1/f_2 \qquad\qquad \therefore \ 1/F = 1/f_1 + 1/f_2$$

Measurement of refractive index by liquid lens method

A convex lens is placed on a plane mirror and the focal length (f_1) determined by no parallax between a pin and its image. A drop of liquid is smeared between the mirror and the lens and the focal length of the combination (F) determined. The radius of curvature r of the underside of the convex lens is determined by Boys's method (page 57). If f_2 is the focal length of the liquid lens

$$\frac{1}{F} = \frac{1}{f_1} + \frac{1}{f_2} \quad \text{(hence } f_2 \text{ may be determined)}$$

and
$$\frac{1}{f_2} = (\mu - 1)\,\frac{1}{r} \quad \text{(lens equation, page 56)}$$

and hence μ may be determined.

Defects of vision

(i) *Short sight* (myopia). The far point is too near to the eye lens. Either the eyeball is too long or the eye lens is too powerful (or both). (ii) *Long sight* (hypermetropia). Close objects cannot be seen clearly. Either the eyeball is too short or the eye lens is too weak (or both). (iii) *Far sight* (presbyopia). A decrease in the range of accommodation frequent in old age, and caused by a weakening of the ciliary muscles. (iv) *Astigmatism.* If objects in one plane are in focus objects in another plane will be blurred. A common cause is when the cornea has different curvatures in different planes.

Microscopes and Telescopes

Definitions

109. The *magnifying power* or *angular magnification of a microscope*

$$= \frac{\text{angle subtended by the final image at the eye}}{\text{angle subtended at the eye by the object placed at the near point}}$$

110. The *magnifying power of a telescope*

$$= \frac{\text{angle subtended by the final image at the eye}}{\text{angle subtended by the object at the eye}}$$

The microscope

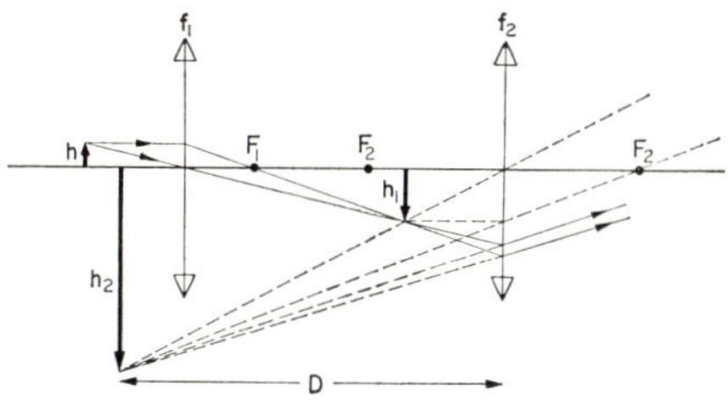

FIG. 68. Microscope in near point adjustment.

If the final image is at the near point D, near point adjustment (Fig. 68), and f_1 and f_2 are the focal lengths of the objective and eyepiece respectively, then

$$\text{Angular magnification} = \frac{h_2/D}{h/D} = \frac{h_2}{h} = \frac{h_2}{h_1} \times \frac{h_1}{h}$$

$$= \text{(linear magnification of eyepiece)}$$
$$\times \text{(linear magnification of objective)}$$

The telescope

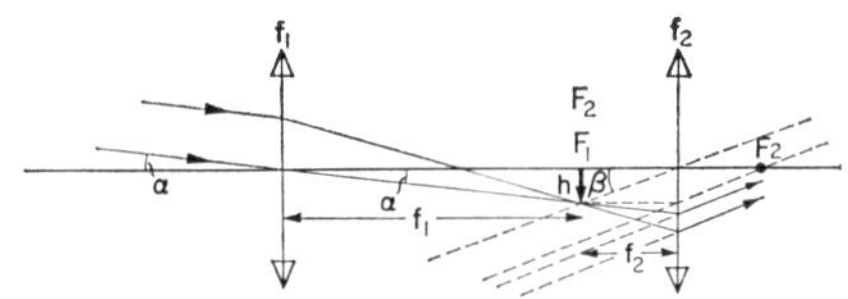

FIG. 69. Telescope in far point adjustment.

If the final image is at infinity, far point adjustment (Fig. 69), then

$$\text{Angular magnification} = \frac{h/f_2}{h/f_1} = \frac{f_1}{f_2}$$

In one type of reflecting telescope (Fig. 70) an image of the object is formed at a hole at the pole of the concave mirror. This image is viewed with an eyepiece. Reflecting telescopes can have

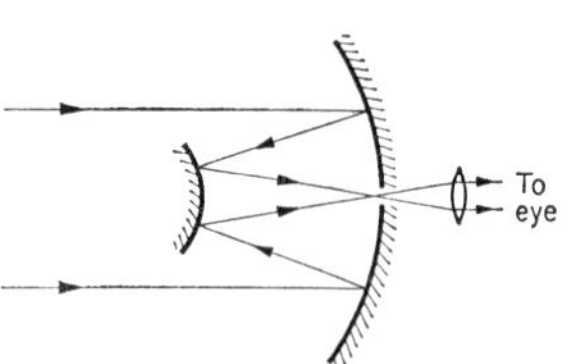

FIG. 70. Reflecting telescope.

larger apertures than lens instruments since mirrors can be supported better and there is no chromatic aberration. Mirrors need fewer optically worked surfaces, and as they are front silvered the material of the mirror does not transmit light. For small telescopes, lenses are used because there is little deterioration of the surface over the years, and there is less obstruction.

The eyering

All the rays which enter the optical system pass through the objective. The light emerging from the eyepiece passes through a ring which is the image of the objective in the eyepiece. This ring is the smallest area through which all the light passes and is known as the eyering. It is the best position for the eye.

Spectrometer and Spectra

The spectrometer (Fig. 71)

To set up a spectrometer the following adjustments are made. (i) Adjust the telescope to receive parallel light by focusing it on a

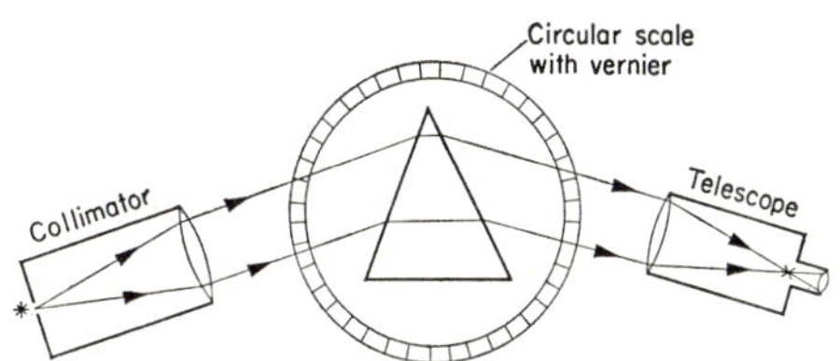

FIG. 71. Spectrometer.

distant object. (ii) Adjust the collimator to give parallel light by moving the slit until its image is in focus on the crosswires in the telescope. (iii) Level the table.

To measure the angle A of a prism. Arrange the prism as indicated in Fig. 72. Measure the angle $2A$ by rotating the telescope.

To determine the refractive index μ. Measure the angle of the prism A and the minimum deviation D (it is more accurate to measure $2D$ and divide by 2). μ is obtained by substitution in eqn. (12), page 51.

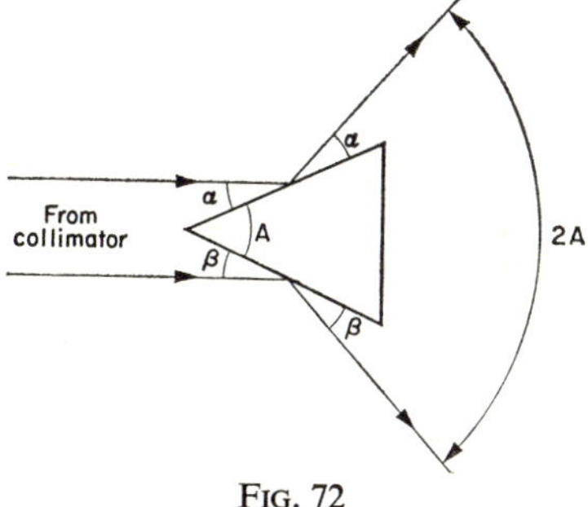

FIG. 72

Classification of spectra

(i) *Emission.* (a) Continuous: emitted by incandescent solids, liquids and gases at high pressure. All wavelengths are present since any vibrations of definite frequency are changed by collisions or interaction between the molecules. (b) Line: emitted by incandescent vapours composed of atoms, e.g. in a bunsen flame, electric arc or discharge tube. The emission is caused by the electrons in the atoms returning to lower energy levels after being excited. (c) Band: emitted by incandescent vapours composed of molecules. Molecules give rise to band spectra as a result of changes in rotational and vibrational energies, e.g. oxygen or nitrogen in a discharge tube.

(ii) *Absorption.* May be line, band or continuous. They are caused by the absorption of light of a particular wavelength(s), e.g. the Sun's spectra (Fraunhöfer lines). Blood gives dark bands in the orange and yellow regions and absorbs all violet light.

Velocity of Light

Michelson's method

A is an eight-sided mirror (Fig. 73). Light from the source S is reflected via the path A, m_1, m_2, B, C, m_3, C, B, m_4, m_5, A, m_6 and then to a microscope at D. The distance from A to C was

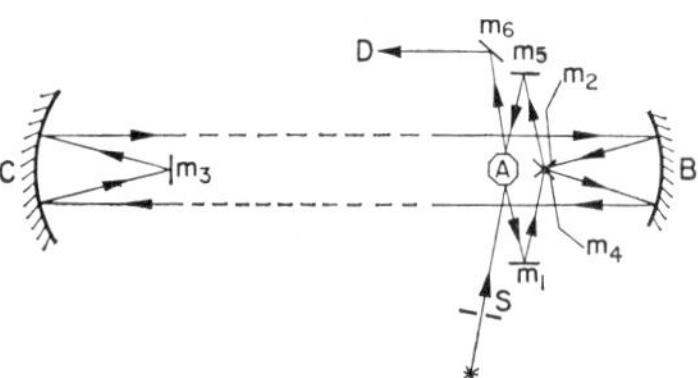

FIG. 73. Michelson's method for the velocity of light.

about 22 miles. The mirror A is rotated until the image of S is stationary on the crosswires of the microscope. When this happens, if d centimetres is the distance the light travels, and n the number of revolutions of the mirror every second then

$$\text{velocity of light} = \frac{\text{distance travelled}}{\text{time taken}} = \frac{d}{1/8n}$$

The method is more accurate than many earlier methods in which it was necessary to judge when the intensity of the light was a minimum. In this method the image will move slowly across the field of view until the speed of the mirror is correctly adjusted. In more recent experiments a similar method was employed to measure the velocity of light in an evacuated pipe, where the temperature and pressure could be measured with greater accuracy than in the open air.

CHAPTER 4

Wave Motion in Light and Sound

Oscillations and Waves

Definitions

111. *Oscillations* are variations in the magnitude of a physical quantity which repeat themselves at regular intervals.

112. A *wave* is an oscillatory disturbance which travels with a finite velocity and continuously transfers energy through a medium (or space) without any movement of the medium as a whole.

113. The *amplitude* of an oscillatory disturbance is the maximum deviation of the varying physical quantity from its mean value.

114. The *frequency* of an oscillatory disturbance is the number of oscillations occurring in unit time (usually hertz (Hz) or cycles/s).

115. Oscillatory disturbances at two points are in *phase* if they are at the same stage of oscillation.

116. A *wavefront* is a continuous line or surface on which all particles are vibrating in the same phase.

117. The *wavelength* is the smallest separation of two points in the path of a wave at which the oscillations are in phase.

118. In a *transverse wave* the oscillations in the varying physical quantity occur in a plane perpendicular to the direction of propagation of the wave.

119. In a *longitudinal wave* the oscillations in the varying physical quantity occur in a plane containing the direction of propagation of the wave.

120. A *harmonic* is an oscillation whose frequency is an exact multiple of the fundamental. The fundamental is the first harmonic.

121. *Overtones* are frequencies, other than the fundamental, actually present in a given sound.

122. *Huygens's construction.* Each small element of a wavefront may be treated as a new source of secondary waves and the position of the whole wavefront at any subsequent time is given by the envelope of the secondary wavelets emitted by the secondary sources.

Many of the phenomena associated with a wave motion may be demonstrated using a ripple tank.

Interference, diffraction and polarization may be explained by a wave theory; polarization requires that the waves be transverse.

Interference in sound

Demonstrations. (i) Rotate a tuning fork near the ear. (ii) Put two small speakers near to one another and listen to a pure note as you walk along a line parallel to the line joining the two speakers. (iii) Beats. If the frequencies of two tuning forks are f_1 and f_2 and t is the time between successive beats, then $f_1 t - f_2 t = 1$ or $f_1 - f_2 = 1/t =$ the beat frequency.

Interference in light

(i) *Young's slits.* Consider the light arriving at P (Fig. 74) from two coherent sources S_1 and S_2.

$$S_2 P^2 - S_1 P^2 = [(x+s/2)^2 + D^2]$$
$$- [(x-s/2)^2 + D^2] = 2xs$$

where s is the slit separation, D the distance from the plane of the sources to the plane where the fringes are being viewed and x the

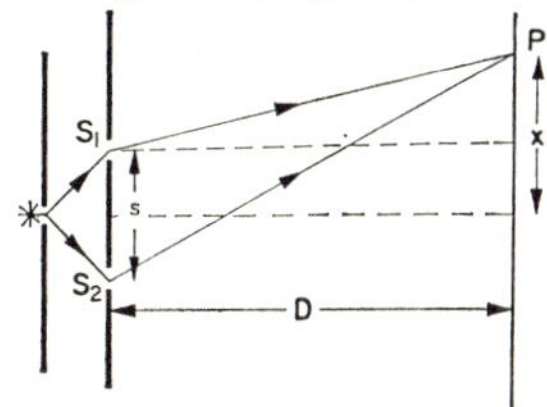

FIG. 74. Young's slits.

distance from the central fringe to the point P.

$$\therefore \; S_2P - S_1P = \frac{2xs}{(S_2P + S_1P)} \simeq \frac{xs}{D} \quad (\text{since } S_2P + S_1P \simeq 2D)$$

For constructive interference $n\lambda = xs/D$, for a dark fringe $(2n+1)\lambda/2 = xs/D$.

With white light the central fringe is white with red on either side (the red fringes are the broadest because λ is largest for red light). Overlapping of orders soon occurs and uniform illumination results.

Experimental details. The slits must be parallel, narrow and close together. Use a travelling microscope with a vernier scale to measure the fringe separation. Measure from the slits to any point on the microscope. Focus the microscope on the slits and measure the slit separation. Measure from the slits to the same point on the microscope as before (the difference from the previous reading is D).

(ii) *Fresnel's biprism.* Two coherent sources may also be produced by using a biprism of small angle (Fig. 75). The slit separation s may be determined by placing a lens at A and producing images of the slits on the crosswires of a fixed microscope. If h_1 and h_2 are the separation of the slit images for the two possible positions of the lens, then $s = \sqrt{(h_1 h_2)}$.

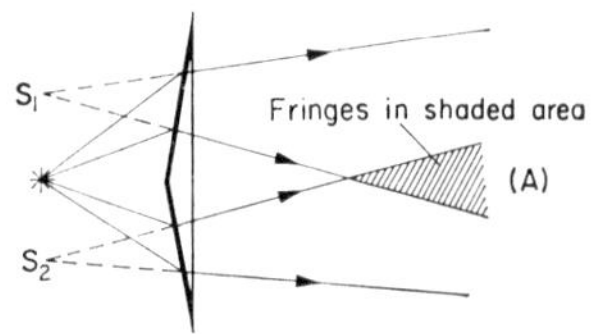

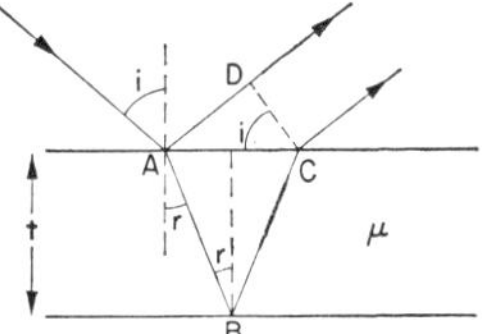

FIG. 75. Fresnel's biprism experiment.   FIG. 76. Interference at thin films.

(iii) *Thin films.* Referring to Fig. 76,

optical path difference $= \mu(AB + BC) - AD = \mu 2t/\cos r - AC \sin i$

$$= 2\mu t/\cos r - 2t \,.\, \tan r \,.\, \sin i$$

Simplifying and using $\sin i = \mu \sin r$ we get:

$$\text{optical path difference} = 2\mu t \cos r$$

Allowing for the phase change when the light is reflected at A, for constructive interference $2\mu t \cos r = (n-\tfrac{1}{2})\lambda$.

With white light some wavelengths interfere constructively and others destructively giving rise to coloured fringes.

(iv) *Newton's rings.* From the geometry of Fig. 77,

$$\rho^2 = (\rho - t)^2 + r^2$$
$$\therefore \; r^2 = \rho 2t - t^2 \simeq \rho 2t$$

For a dark fringe $r_n^2 = \rho n\lambda$. A graph of r_n^2 against n is a straight line through the origin with gradient $\rho\lambda$. When there is optical

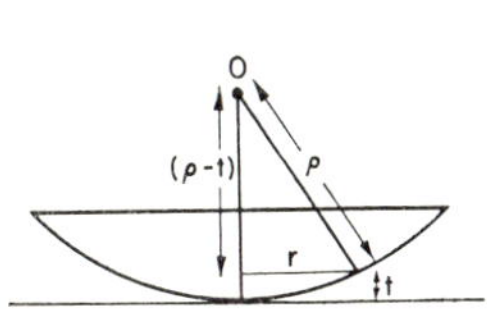
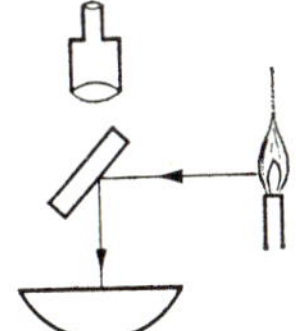

FIG. 77. Newton's rings.

contact at the centre, the central spot is dark because of a phase change of π at any air/glass reflection. The value of ρ must be large otherwise the fringe separation is too small to be visible.

Conditions for visible interference

(i) Coherent sources, i.e. the same frequency and a constant phase relationship (synchronized phase changes). (ii) Sources must be close together. (iii) Path difference must be small or there will be no fringes due to random phase changes between bursts of waves. (iv) Same amplitude for complete destructive interference.

Diffraction grating (Fig. 78)

Consider the light which is diffracted through an angle θ when parallel light falls normally on the grating. If e is the slit separation, then $e \sin \theta = n\lambda$ for a bright fringe. The diffracted light is usually viewed through a spectrometer telescope which brings different parts of the plane wavefront to a focus. With white light a continuous spectrum is produced.

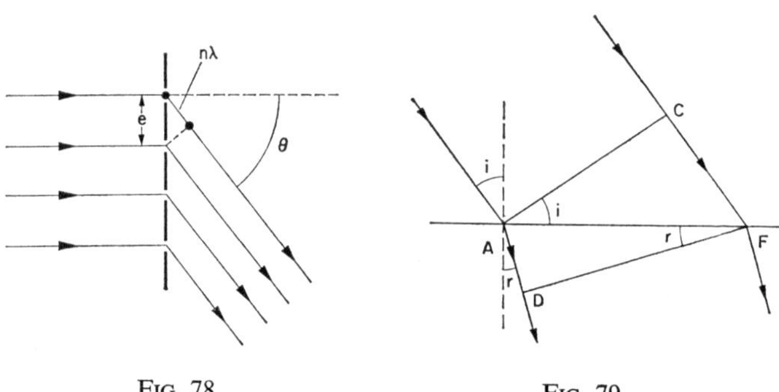

FIG. 78 FIG. 79

Refractive index and velocity of light

The wavefront AC (Fig. 79) is refracted along DF. Let V_m be velocity in the medium and V_a the velocity in air. If t is the time taken to travel the distances CF and AD, then

$$AD = V_m t \quad \text{and} \quad CF = V_a t$$

But

$$\mu = \frac{\sin i}{\sin r} = \frac{CF/AF}{AD/AF} = \frac{CF}{AD} = \frac{V_a}{V_m}$$

Polarization

Light is a transverse electromagnetic wave. The energy is propagated away from the source by varying electric and magnetic fields (E and H). In a plane polarized wave E and H are perpendicular to each other and each is confined to a particular direction in a plane perpendicular to the direction of travel of the light.

Production. (i) Pass light through a crystal of tourmaline. The emergent light is plane polarized along the axis of the crystal. (ii) Double refraction. Pass the light through a crystal of iceland spar. Two beams are produced plane polarized perpendicular to each other. (iii) Reflect light from crown glass at an angle of incidence of 57°. The light is plane polarized parallel to the surface. (iv) Use a sheet of polaroid. This consists of a sheet of thin polyvinyl alcohol stretched to arrange the molecules in long parallel chains.

Brewster's law. At the polarizing angle (57° for crown glass) the reflected and refracted rays are 90° apart.

$$\sin i / \sin (90 - i) = \mu = \tan i$$

Detection. View the light through either a tourmaline crystal or a sheet of polaroid. As the detector is rotated there will be a position when no light passes if the light is totally plane polarized.

Applications. (i) Determination of concentration of optically active organic substances. The amount of rotation of the plane of polarization is a measure of the concentration. (ii) Ophthalmoscope. The disturbing reflections from the cornea are eliminated. The retina remains visible because it depolarizes the light. (iii) Filters for cameras. (iv) Elimination of glare from reflected light by using polaroid sun glasses. (v) Anti-dazzle windscreens and headlamps.

To show $v = f\lambda$

If a source is emitting a note of wavelength λ and frequency f, in one second each crest travels a distance $f\lambda$. If v is the velocity of the wave then $v = f\lambda$.

THE ELECTROMAGNETIC SPECTRUM

Name	Produced by	Approximate range
Radio	Electronic oscillating circuits	$5.10^5\text{cm}{-}10^{-1}\text{cm}$
Infrared	Hot bodies (oscillation of molecules)	$10^{-1}\text{cm}{-}8.10^{-5}\text{cm}$
Visible light	Hot bodies (atomic oscillations)	$8.10^{-5}\text{cm}{-}4.10^{-5}\text{cm}$
Ultraviolet	Atomic oscillations (orbital electrons)	$4.10^{-5}\text{cm}{-}10^{-7}\text{cm}$
X-rays	Atomic oscillations (innermost orbital electrons)	$10^{-7}\text{cm}{-}10^{-10}\text{cm}$
Gamma rays	Oscillations of nucleus (disposal of surplus energy)	$10^{-8}\text{cm}{-}10^{-11}\text{cm}$

Standing waves

A standing wave results from the interference of two progressive waves of equal frequency and amplitude travelling in opposite directions. (i) Some points, nodes, are permanently at rest. (ii) Points between successive nodes are vibrating with the same frequency and phase, the amplitude of the vibration increasing between node and antinode. (iii) The distance between successive nodes is $\lambda/2$. (iv) In sound a displacement antinode is a pressure node, and a displacement node is a pressure antinode. In contrast a plane progressive wave is one which causes the same amplitude at all points and neighbouring points vibrate out of phase.

Resonance

Resonance is a term to describe the situation in which oscillations of large amplitude are built up by the application of impulses of small amplitude to a system having a natural frequency equal to the frequency of the applied impulses. Examples are a tuning fork held over a flask containing the correct amount of water, a sonometer wire being set in vibration by a tuning fork held on the sonometer board, and the vibration of part of the bodywork of a motor car when the engine is running at a particular speed.

Resonance tube experiment to determine the velocity of sound

The water level is lowered until resonance occurs when tuning forks of known frequency f are held above the tube. If l is the length of the tube at resonance, then $\lambda/4 = l + c$ where c is the end correction. But $v = f\lambda$, therefore $l + c = v/4f$ where v is the velocity of sound. A graph of l against $1/f$ is a straight line of gradient $v/4$.

Modes of vibration of a closed pipe (Fig. 80)

The closed end must be a displacement node since the air molecules cannot move. The open end must be a pressure node (a displacement antinode) because the pressure is always atmospheric.

Fundamental (1st harmonic); $l = \lambda/4$, frequency f_0.
1st overtone (3rd harmonic); $l = 3\lambda/4$, frequency $3f_0$.
2nd overtone (5th harmonic); $l = 5\lambda/4$, frequency $5f_0$.

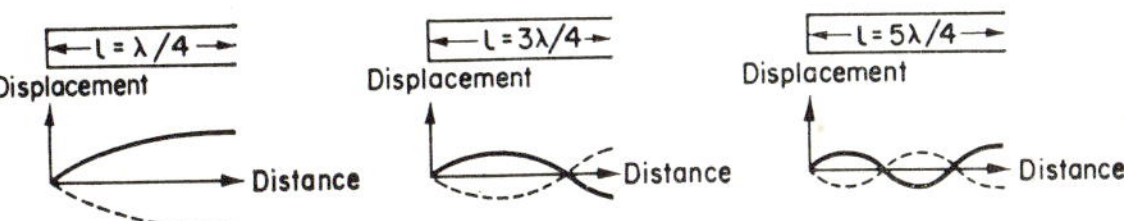

FIG. 80. Modes of vibration of a closed pipe.

Modes of vibration of an open pipe (Fig. 81)

Fundamental (1st harmonic); $l = \lambda/2$ and $f_0 = v/2l$
1st overtone (2nd harmonic); $l = \lambda$ and $f_1 = v/l = 2f_0$
2nd overtone (3rd harmonic); $l = 3\lambda/2$ and $f_2 = 3f_0$.

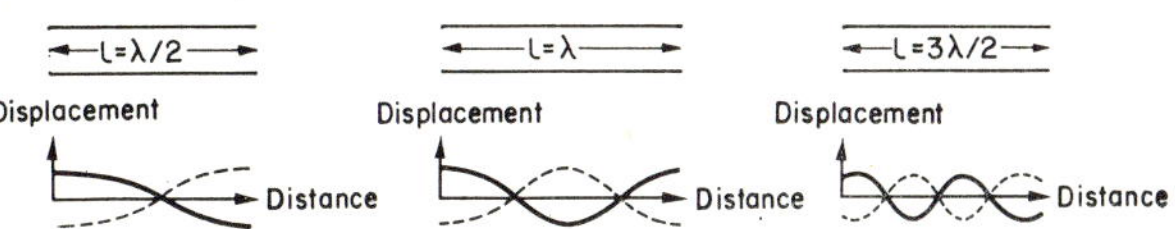

FIG. 81. Modes of vibration of an open pipe.

Vibrations in strings

For the fundamental (Fig. 82) there are nodes N at the ends and an antinode A in the middle. $l = \lambda/2$ and $f_0 = v/2l = \frac{1}{2l}\sqrt{(T/m)}$ since $V = \sqrt{(T/m)}$ where T is the tension and m the mass per unit length.

For the 1st overtone (3rd harmonic) for a string bowed at centre (Fig. 83), $l = 3\lambda/2$ and $f_1 = v/\frac{2}{3}l = 3f_0$. A vibration of frequency $2f_0$ (2nd harmonic) may be obtained if the string is touched lightly at the centre and plucked one-quarter of the way from one end.

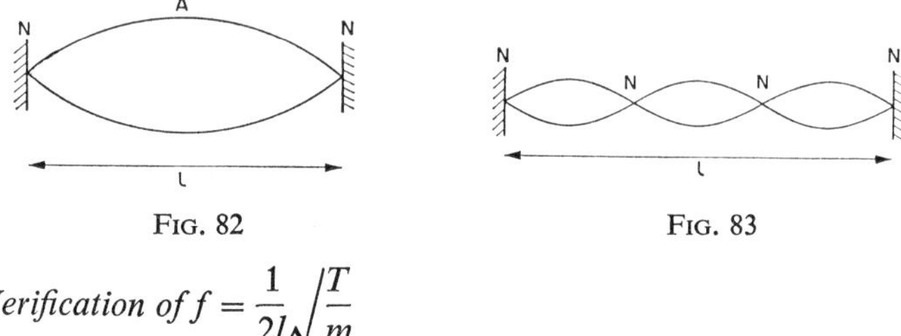

Fig. 82   Fig. 83

Verification of $f = \dfrac{1}{2l}\sqrt{\dfrac{T}{m}}$

Use a sonometer and keep two of the variables constant.

Hebb's method for the velocity of sound in free air (Fig. 84)

M_1 and M_2 are microphones. When the waves reaching the microphones are out of phase the sound in the telephone T will

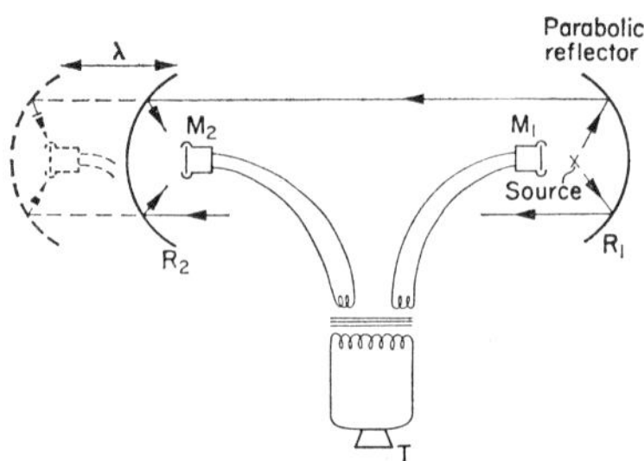

Fig. 84. Hebb's method for the velocity of sound.

be a minimum. The distance the reflector R_2 and microphone M_2 are moved between successive minima is equal to the wavelength λ of the sound. The frequency of the source f is known and v is calculated from the formula $v = f\lambda$.

Velocity of sound in a gas

$$v = \sqrt{\left(\frac{\gamma p}{\rho}\right)} \tag{15}$$

where p is the pressure, ρ the density and γ the ratio of the specific heat at constant pressure to the specific heat at constant volume. Since $PV/T = A$, where A is a constant for a given mass of gas, and $\rho = M/V$, where M is the mass of volume V,

$$v = \sqrt{\left(\frac{\gamma A T}{M}\right)} \tag{16}$$

   (i) The velocity is independent of the pressure (in eqn. (15) if p doubles, then ρ doubles).

  (ii) The velocity is proportional to the square root of the absolute temperature.

 (iii) The velocity is slightly greater in moist air than in dry air. This is because the density of air containing unsaturated water vapour is less than the density of dry air. This is partly offset by the fact that γ for unsaturated water vapour is less than γ for oxygen and nitrogen.

Kundt's tube (Fig. 85)

   (i) *To demonstrate standing waves.* When the rod which is clamped at its midpoint is stroked with a rosined cloth it vibrates

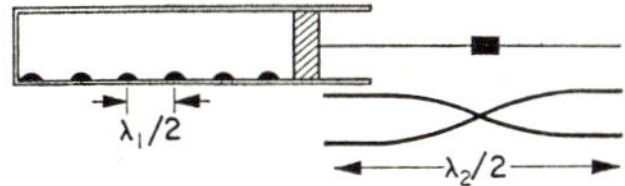

FIG. 85. Kundt's tube.

longitudinally and emits a high pitched note. When the length of the glass tube is correctly adjusted the lycopodium powder in the tube can be observed in heaps at the nodes and in ridges at the antinodes. The distance between two nodes is $\lambda_1/2$ where λ_1 is the wavelength of the note in air.

(ii) *To determine the velocity of sound in the rod.* Let v_1 and v_2 be the velocities of sound in air and the rod respectively. The wavelength of the sound in the rod, λ_2, is twice the length of the rod since the centre of the rod must be a node. We have

$$v_1 = f\lambda_1 \quad \text{and} \quad v_2 = f\lambda_2 \qquad \therefore \ v_2 = \frac{\lambda_2}{\lambda_1}v_1$$

Doppler Effect

Suppose a source S (Fig. 86) is emitting a note of frequency f and moving towards an observer O with a velocity u_s. Suppose also that the observer is moving away from the source with a

FIG. 86

velocity u_o. If the velocity of the wave motion is V then there will be f maxima in a distance $(V-u_s)$. The maxima in a distance $(V-u_o)$ pass the observer's ear every second.

In a distance $(V-u_s)$ there are f maxima.

In a distance $(V-u_o)$ there are $\dfrac{f}{(V-u_s)} \cdot (V-u_o)$ maxima.

$$\text{Apparent frequency} = \frac{(V-u_o)}{(V-u_s)}f$$

The increase in pitch of the horn of an approaching motor-car is an example of the Doppler effect. The effect is also observable in light; the spectrum of a receding star being shifted towards the red end when compared with a terrestrial source.

Characteristics of Notes

Definitions

123. *Pitch.* Pitch is the characteristic of a tone by which the ears assign it a place in the musical scale. Pitch is primarily determined by the frequency.

124. *Quality.* The tone quality or timbre of a note is determined by the relative strengths of the various overtones present.

125. *Intensity.* The intensity is the power transmitted through unit area and it is usually measured in microwatts per square metre.

126. *Relative intensity.* When the intensity of one sound is 10 times that of another, its sound level is said to be 1 *bel* higher than the other, or

$$\text{number of decibels} = 10 \log_{10}(I_2/I_1)$$

where I_1 is the original power and I_2 the final power.

127. *Loudness.* How loud a sound appears depends on the ear of the observer. The unit of loudness is the *phon*. It is numerically equal to the relative intensity in decibels above the threshold value ($10^{-18}\ \mu\text{W m}^{-2}$) of a standard note of 1000 H_2 which "the average person" judges to be equally loud.

CHAPTER 5

Electricity and Magnetism

Basic Principles

Definitions and laws

128. One *ampere* (A) is that current which, flowing in both of two infinitely long straight parallel conductors of negligible cross-section 1 m apart *in vacuo*, produces between them a force of 2×10^{-7} N m^{-1} of their length.

129. One *coulomb* (C) is the charge which flows past a point in 1 s when a current of 1 A is flowing.

130. A p.d. of 1 *volt* (V) is the potential difference between two points when 1 J of work is done in moving 1 C between them.

131. *Ohm's law.* The current flowing through a metallic conductor is proportional to the potential difference across it provided that the temperature remains constant.

132. One *ohm* (Ω) is the resistance of a conductor when 1 V across its ends causes a current of 1 A to flow.

133. When a current flows parallel to the axis of a uniform cylinder of length l and cross-sectional area A, the resistance R at a temperature T is given by $R = \rho l / A$ where ρ is the *resistivity* of the material of the conductor at temperature T (usually Ω m or Ω cm).

134. A *watt* (W) is a rate of working of 1 J s^{-1}.

135. A *kilowatt hour* (kWh) is the energy transformed in 1 hr at a rate of working of 1000 W; it is equal to 3,600,000 J.

76

136. The *electromotive force* (e.m.f.) of a source of electrical energy is the energy it supplies when 1 C of electricity is passed through it. It is the p.d. across the terminals when in open circuit (usually $J\,C^{-1}$ or V).

137. The *temperature coefficient of resistance* (α) of a material is the increase in resistance of unit resistance at 0°C for a 1°C rise in temperature (unit: per °C).

138. *Kirchhoff's laws.* (i) The sum of the currents arriving at any junction is equal to the sum of the currents leaving that junction.

(ii) For any closed circuit the algebraic sum of the p.d. drops is equal to the algebraic sum of the e.m.f.'s.

The heating effect of a current

From the definition of a volt, 1 joule of work is done when 1 coulomb is moved through a p.d. of 1 volt. QV joules of work are done when Q coulombs are moved through a p.d. of V volts.

Suppose the Q coulombs are carried by I amps flowing for t seconds, then $Q = It$, and the work done $= VIt$ joules. Using Ohm's law we have

$$\text{work done} = VIt = I^2Rt = V^2t/R \text{ joules}$$

Wheatstone (and metre) bridge (Fig. 87)

If *no current flows in the galvanometer* and if I_1 is the current in ABD and I_2 is the current in ACD, then p.d. between A and B = p.d. between A and C,

i.e. $\qquad I_1R_1 = I_2R_3$

p.d. between B and D

$\qquad\qquad$ = p.d. between C and D,

i.e. $\qquad I_1R_2 = I_2R_4$

Dividing: $\underline{R_1/R_2 = R_3/R_4}$ (17)

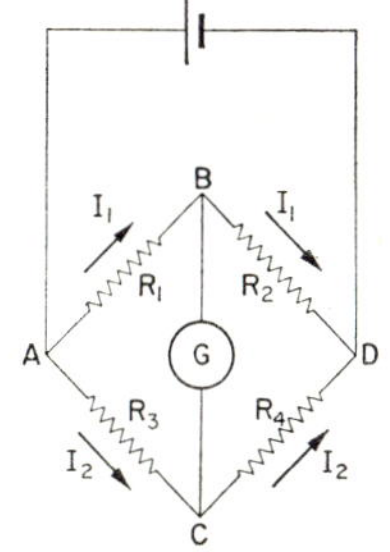

FIG. 87. Wheatstone bridge.

If *ACD* is a metre wire (Fig. 88) of uniform resistance, then its resistance will be proportional to its length and

$$R_1/R_2 = l_1/l_2$$

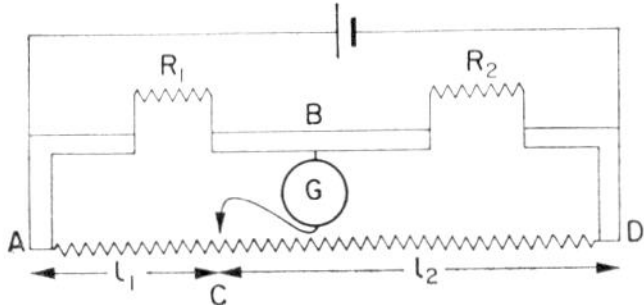

FIG. 88. Metre bridge.

Temperature coefficient of resistance

If R_t is the resistance at $t°C$ and R_0 the resistance at $0°C$ then from definition 137

$$\alpha = \frac{R_t - R_0}{R_0 t} \qquad \therefore R_t = R_0\,\alpha t + R_0$$
$$\therefore R_t = R_0\,(1 + \alpha t)$$

R_t may be measured using a metre bridge, and if a graph is plotted of R_t against t, then the intercept when $t = 0$ is R_0 and the gradient is $R_0\alpha$. Hence α may be determined.

The Potentiometer

A potentiometer consists of a cell of constant e.m.f. and a wire whose resistance per unit length is constant. It is an instrument for comparing two p.d.'s.

Comparison of e.m.f.'s (Fig. 89)

With the switch S closed, if no current flows in the galvanometer, the p.d. across the terminals of the cell of e.m.f. E_1 is equal to the p.d. across l_1

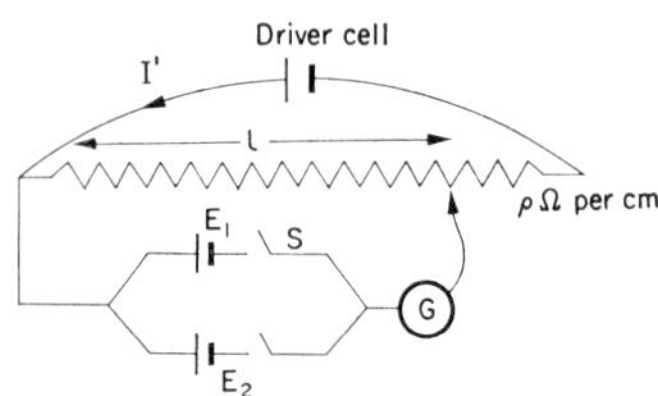

FIG. 89. Comparison of e.m.f.'s.

centimetres of wire, i.e. $E_1 = I'\rho l_1$ where I' is the current flowing in the wire, and ρ the resistance of the wire per centimetre. Similarly, if the e.m.f. E_2 is balanced by the p.d. drop across l_2 centimetres of wire, then $E_2 = I'\rho l_2$. Dividing

$$\frac{E_1}{E_2} = \frac{I'\rho l_1}{I'\rho l_2} = \frac{l_1}{l_2}$$

Measurement of current (or calibration of an ammeter)

The p.d. across the ends of a standard resistance R_{st} (Fig. 90) is compared to the e.m.f. of a standard cell E_{st}. Using the notation

FIG. 90. Calibration of ammeter.

indicated in the diagram

$$IR_{st} = I'\rho l \quad \text{and} \quad E_{st} = I'\rho l_{st}$$

Dividing $\qquad\qquad I = E_{st} \cdot l / R_{st} \cdot l_{st}$

Comparison of resistances

The balance lengths l_1 and l_2 (Fig. 91) are found for the *same* current flowing through R_1 and R_2. At balance $IR_1 = I'\rho l_1$ and $IR_2 = I'\rho l_2$. Hence

$$R_1/R_2 = l_1/l_2$$

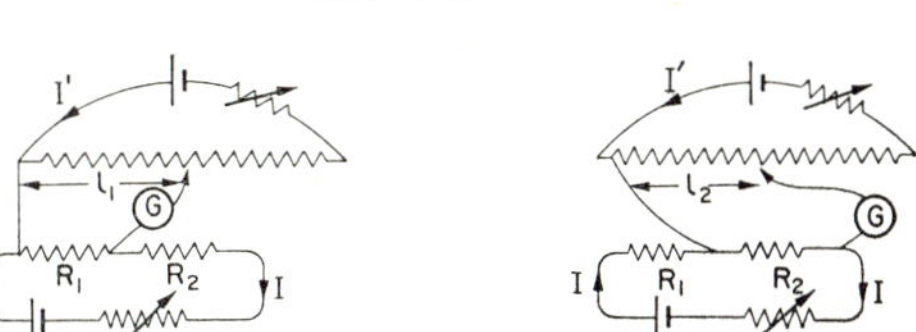

FIG. 91. Comparison of resistances.

D

The variable resistances enable a series of readings to be taken. The resistance in the driver cell circuit is adjusted so that a high balance is obtained thus ensuring greater accuracy. The method is particularly suitable for use with four terminal low resistors.

Internal resistance of a cell

For the circuit (Fig. 92) containing the cell of e.m.f. E of internal resistance r with a resistance R across its terminals, $E = I(R+r)$ and the p.d. across the terminals $V = IR$. Dividing

$$\frac{E}{V} = \frac{R+r}{R} = \frac{l'}{l}$$

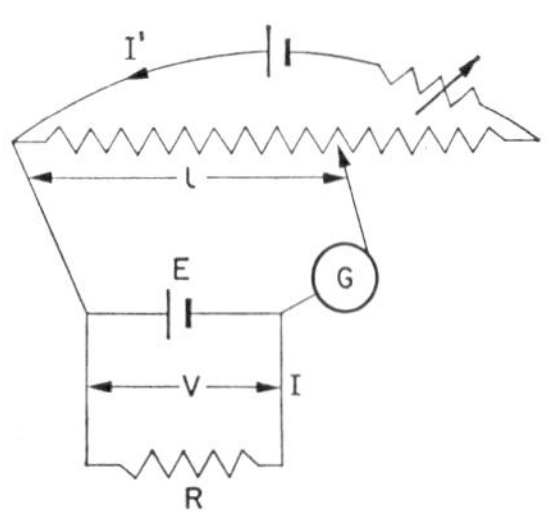

FIG. 92. Internal resistance of a cell.

where l' is the balance length when no current flows through the cell, and l is the balance length for a given resistance R. Hence if R is known r may be calculated. If a series of readings of R and l are made, then a graph of $1/R$ against $1/l$ will have an intercept on the $1/R$ axis equal to $-1/r$.

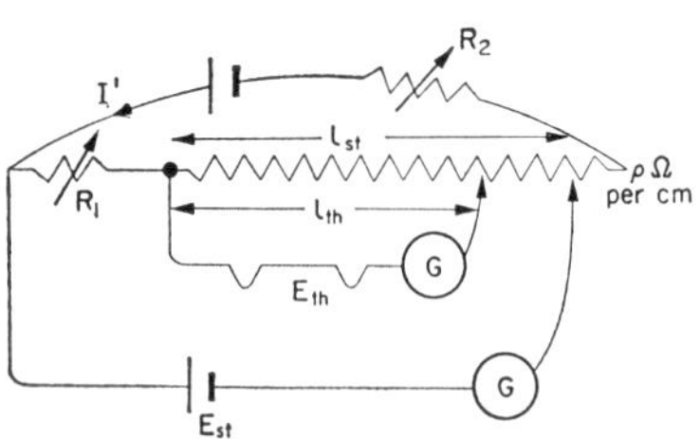

FIG. 93. E.m.f. of a thermocouple.

Measurement of small p.d.'s

The circuit is shown in Fig. 93. A large resistance R is in series with the wire, and the p.d. across the wire is small. $(R_1 + R_2)$ is kept constant. If E_{th} is the e.m.f. of the thermocouple, then using the notation shown in Fig. 93,

$$E_{th} = I'\rho l_{th} \quad \text{and} \quad E_{st} = I'(R_1 + \rho l_{st})$$

Dividing $\quad E_{th} = \dfrac{\rho l_{th} E_{st}}{R_1 + \rho l_{st}}$

In all potentiometer experiments the balance point must be as large as possible if a high degree of accuracy is required. The readings should be repeated to check the constancy of the e.m.f. of the driver cell.

Determination of resistivity

Measure the resistance R using a metre bridge and substitute in the equation $R = \rho l/A$.

Resistances in series

$$R = R_1 + R_2 + R_3.$$

Resistances in parallel

$$1/R = 1/R_1 + 1/R_2 + 1/R_3.$$

Magnetism and Electromagnetism
(m.k.s. units. Even-numbered pages 82, 84, 86, 88)

Definitions

139. The *magnetic flux* (ϕ) through a surface is the time integral of the e.m.f. induced in a conductor of negligible cross-sectional area which is moving over the surface [unit: volt.second (Vs) or *webers* (Wb)].

140. The *magnetizing force* (H) inside a long solenoid is the linear surface density of current (number of ampere turns per metre) flowing in it [unit: ampere (turns) per metre (A m^{-1})]. The magnetizing force at any point is the linear surface density of current which, flowing in an infinite solenoid, would produce the same magnetizing force inside the solenoid as exists at the point.

141. The *magnetic flux density* (B) is the magnetic flux which passes normally through unit area, the area being normal to the direction of the flux [unit: tesla (T) = Wb m^{-2}].

142. The *permeability* (μ) of a medium is the ratio of the magnetic flux density to the magnetizing force (unit: Wb A^{-1} m^{-1}).

143. The *relative permeability* (μ_r) is the ratio of the permeability of the medium to that of free space.

Mathematical summary of above definitions

$$\phi = \int e \, . \, \mathrm{d}t = BA, \quad \mu = B/H, \quad \mu_r = \mu/\mu_0, \quad H = NI/L$$

where μ_0 is the permeability of free space, N the number of turns on a solenoid of length L carrying a current I.

Magnetism and Electromagnetism
(c.g.s. units. Odd-numbered pages 83, 85, 87, 89)

Definitions

144. The force between two magnetic point poles is directly proportional to the product of their pole strengths and inversely proportional to the square of the distance between them. It also depends on the medium

$$F = \frac{m_1 m_2}{\mu d^2} \text{ dynes}$$

where μ is the permeability of the medium. For a vacuum $\mu = 1$.

145. *Unit magnetic pole* repels an identical pole 1 cm away in a vacuum with a force of 1 dyn.

146. A *magnetic field* exists in any region in which a magnetic pole experiences a force.

147. The *magnetic field strength* at a point is the force in dynes on unit N-pole placed in the field at the point (usually oersteds).

148. A *line of magnetic force* is a line whose direction at any point represents the direction of the resultant magnetic field at the point. Lines of force only exist in a vacuum; in any other medium they become lines of induction.

149. The *magnetic moment* of a bar magnet is the restoring couple on it when it is at right angles to a field of strength 1 oersted.

150. The *flux* linking a circuit is the total number of lines of induction threading the circuit (usually maxwells).

151. The *flux density* (B) is the number of lines of induction per square centimetre normal to the surface (usually gauss).

152. The *permeability* of a medium is the ratio of the flux density to the field strength ($\mu = B/H$).

153. An *e.m.u. of current* is that current which when flowing in a circle of 1 cm radius produces a field of 2π oersteds at the centre.

(m.k.s. units)

Flux density due to a current element

$$\delta B = \frac{\mu I \delta l \sin \theta}{4\pi r^2}$$

This is known as Biot and Savart's law. δB is the flux density (Wb m^{-2}) due to a current element of length δl (m) at a distance r (m) from the element carrying a current I (A) and making an angle θ with it. It cannot be verified directly but its "proof" lies in the many verifiable deductions which can be made from it.

Flux density due to a long straight wire

Flux density at P (Fig. 94) due to current element is

$$\delta B = \frac{\mu I \delta l \cos \phi}{4\pi r^2}$$

But $l = x \tan \phi$

$\therefore \ \delta l = x \sec^2\phi \ d\phi$

and

$$r = x \sec \phi$$

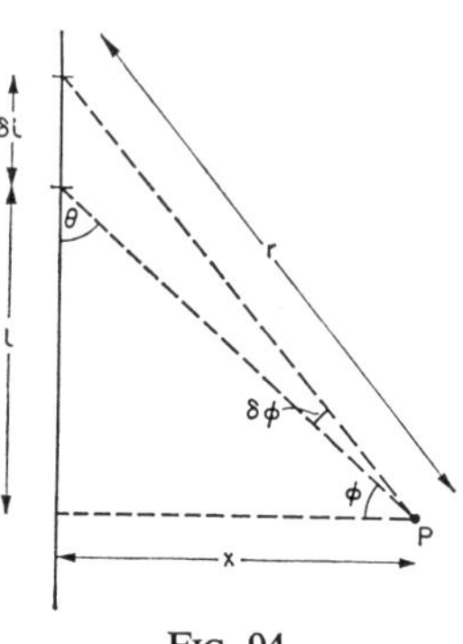

FIG. 94

$$\therefore \ B = \int_{-\pi/2}^{+\pi/2} \frac{\mu I x \sec^2\phi \ d\phi . \cos \phi}{4\pi(x^2 \sec^2\phi)} = \frac{\mu I}{4\pi x} [\sin \phi]_{-\pi/2}^{+\pi/2} = \frac{\mu I}{2\pi x}$$

Flux density on the axis of a circular coil

The flux density at P due to the current element may be resolved into two directions as shown (Fig. 95). The sum of the components

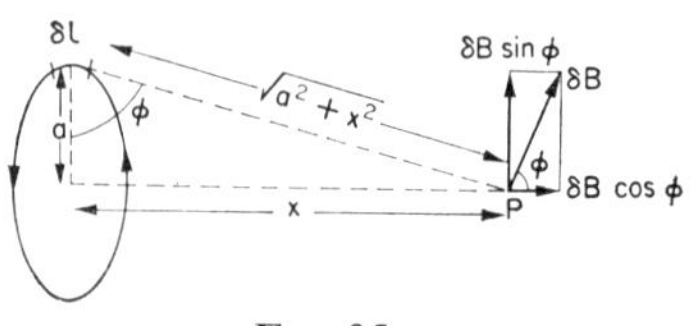

FIG. 95

(c.g.s. units)

154. An *e.m.u. of charge* (or abcoulomb) is that charge which flows past a point in 1 sec when a current of 1 abamp is flowing (= 10 C).

155. A *p.d. of* 1 *e.m.u.* (or abvolt) is the potential difference between two points when 1 erg of work is done in moving 1 abcoulomb between them.

156. An *e.m.u. of resistance* (or abohm) is the resistance of a conductor when 1 abvolt across its ends causes a current of 1 abamp to flow.

Couple on a magnetic dipole in a magnetic field

If m is the pole strength, M the magnetic moment, $2l$ the magnetic length of the magnet and θ the angle it makes with a field of strength H oersteds, then the couple C (Fig. 96) is

$$mH \, . \, 2l \sin \theta$$

and since the magnetic moment M is $m \, . \, 2l$

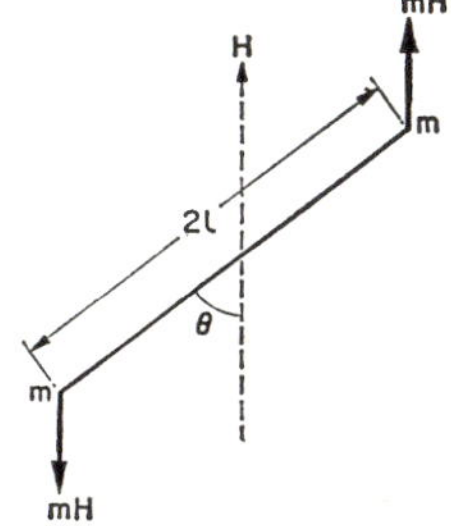

Fig. 96

$$\underline{\text{Couple} = MH \sin \theta}$$

Field strength due to a current element

$$\delta H = \frac{I \delta s \sin \theta}{r^2}$$

This is known as Biot and Savart's law. δH is the field strength due to a current element of length δs at a distance r from the element carrying a current I and making an angle θ with it. The "proof" of this formula lies in the many verifiable deductions which can be made from it.

(m.k.s. units)

$\delta B \sin \phi$ for the whole coil is zero. Hence flux density (B) at

$$P = \sum \delta B \cos \phi = \sum \frac{\mu I \delta l}{4\pi(a^2 + x^2)} \cdot \frac{a}{\sqrt{(a^2 + x^2)}}$$

since θ is 90°.

$$\therefore \; B = \frac{\mu I \cdot 2\pi a N \cdot a}{4\pi(a^2 + x^2)\sqrt{(a^2 + x^2)}} = \frac{\mu I N a^2}{2(a^2 + x^2)^{3/2}}$$

where N is the number of turns and a the radius of the coil.

Flux density at the centre of a circular coil

In the above equation x is zero and $B = \mu I N/2a$.

Flux density inside a toroid

A toroid approximates very closely to an infinitely long solenoid and the flux density B is given by $B = \mu NI/L$, where N is the number of turns, I the current flowing and L the length of the toroid.

Force on a current-carrying conductor in a magnetic field

The conductor (Fig. 97) is moved a distance δx along two parallel conductors. Let B be the flux density perpendicular to the plane of the rails. The induced e.m.f. E is $Bl\delta x/\delta t$. If this

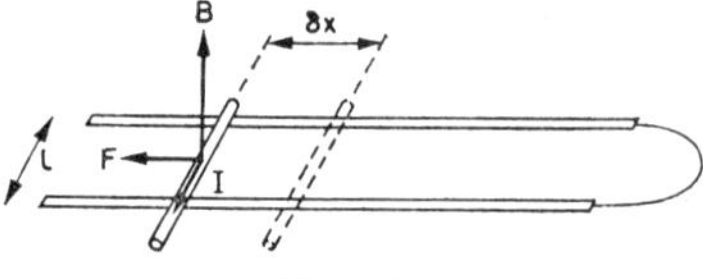

FIG. 97

e.m.f. induces a current I, the heat produced in the wire is $EI\delta t = Bl\delta x \cdot I$. This is equal to the work done $F\delta x$. Therefore

$$F\delta x = Bl\delta x I \quad \text{or} \quad \underline{F = BIl}$$

(c.g.s. units)

Field due to a long straight wire

Field at P due to the current element (Fig. 94) is

$$\delta H = \frac{I\delta l \cos \phi}{r^2}$$

But $l = x \tan \phi$.

$$\therefore \quad \delta l = x \sec^2\phi \, d\phi \quad \text{and} \quad r = x \sec \phi$$

$$\therefore \quad H = \int_{-\pi/2}^{+\pi/2} \frac{Ix \sec^2\phi \cdot d\phi \cdot \cos \phi}{x^2 \sec^2\phi} = \frac{I}{x} \int_{-\pi/2}^{+\pi/2} \cos \phi \, d\phi$$

$$= \frac{I}{x}[\sin \phi]_{-\pi/2}^{+\pi/2} = \underline{\frac{2I}{x}}$$

Field on the axis of a circular coil

The field (B/μ) at P (Fig. 95) due to the current element may be resolved into two directions as shown. The sum of the components $\delta H \sin \phi$ for the whole coil is zero. Hence the field at P

$$= \sum \delta H \cos \phi = \sum \frac{I \cdot \delta l}{(x^2 + a^2)} \cdot \frac{a}{\sqrt{(x^2 + a^2)}}$$

since $\theta = 90°$

$$\therefore \quad H = \frac{I \cdot 2\pi a N}{(x^2 + a^2)} \cdot \frac{a}{\sqrt{(x^2 + a^2)}}$$

where N is the number of turns and a the radius of the coil.

$$\therefore \quad H = \frac{2\pi a^2 N I}{(x^2 + a^2)^{3/2}}$$

Field at the centre of a circular coil

In the above equation x is zero and $H = 2\pi N I / a$.

(m.k.s. units)

Force on two long straight wires carrying a current

Let the wires be a distance d apart and carry currents I_1 and I_2. The conductor carrying a current I_2 is situated in a flux density of $\mu I_1/2\pi d$. Hence force on a length l of this conductor is

$$BI_2 l = \frac{\mu I_1}{2\pi d} \cdot I_2 \cdot l$$

For two wires a metre long and a metre apart, both carrying unit current, and situated in a vacuum where $\mu_0 = 4\pi \times 10^{-7}$ Wb A^{-1} m^{-1}, the force per unit length will be 2×10^{-7} N. This number is used to define the ampere (see page 76).

Magnetic moment of a coil and magnetic dipole

For a rectangular coil of N turns, area A, carrying a current I, situated in a radial magnetic field of flux density B, the couple exerted on the coil is given by

$$G = BANI \quad \text{(see pages 90 and 91)}$$

The *magnetic moment* (M) of the coil is defined by the equation

$$M = G/H \text{ weber metre}$$

or by $\quad\quad\quad\quad\quad M = G/B \text{ amp metre}^2$

It is, therefore, the couple the coil experiences when held with its axis at right angles to a field in which the magnetizing force (or magnetic flux density) is unity.

When it is necessary to distinguish between the two, G/H is called the magnetic dipole moment and G/B the magnetic area moment.

A magnetic dipole and a current carrying coil are completely equivalent as far as their external magnetic effects are concerned, and the above equations are also used to define the magnetic moment of a bar magnet.

(c.g.s. units)

Field inside a solenoid

The field at P (Fig. 98) due to a length δx of the solenoid is given by

$$\delta H = \frac{2\pi a^2 (n\delta x)I}{(x^2 + a^2)^{3/2}}$$

But $x = a \cot \phi$. Therefore $dx = -a \operatorname{cosec}^2\phi \, d\phi$ and

$$(x^2 + a^2) = a^2 \operatorname{cosec}^2\phi$$

$$\therefore \ H = \int_{\phi_1}^{\phi_2} \delta H = \int_{\phi_1}^{\phi_2} -2\pi nI \sin \phi \, d\phi = 2\pi nI(\cos \phi_2 - \cos \phi_1)$$

For an infinitely long solenoid $\phi_1 = \pi$ and $\phi_2 = 0$, hence

$$H = 4\pi nI$$

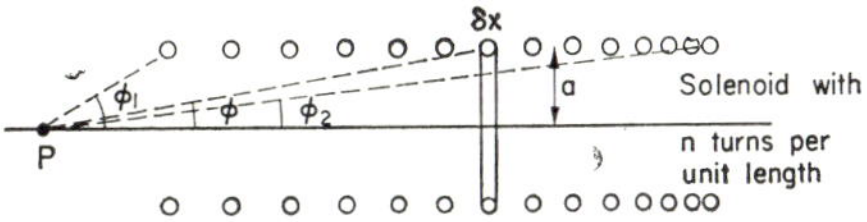

FIG. 98

Force on a wire carrying a current

Consider a coil of one turn and radius a centimetres carrying a current I e.m.u. The field strength at the centre is $2\pi I/a$ oersteds. The force on a pole of strength m at the centre of the circle is $2\pi Im/a$ dynes. By Newton's third law, the force per centimetre on the wire is $2\pi Im/a \div 2\pi a = Im/a^2$. But m/a^2 is the flux density B at the wire due to the pole. Hence the force per centimetre is $I.B$. Therefore the force on a length l of wire is given by

$$\text{Force} = BIl \text{ dynes}$$

Induced e.m.f. in a moving conductor

The conductor is moved a distance δx along two parallel conductors by applying a force to it (a similar arrangement to Fig. 97). Let B be the flux density perpendicular to the rails. An e.m.f. E and current I will be induced in the circuit. By the conservation of energy $F\delta x = EI\delta t$. But $F = BIl$ and the change in flux $\delta\phi = -Bl\delta x$. Substituting we get

$$E = -\frac{d\phi}{dt}$$

Earth's horizontal field (H_0) using a tangent galvanometer

The galvanometer is set with its plane in the direction of the Earth's field. If the magnet rotates through θ when a current I passes through the coil of N turns and radius r, then

c.g.s. units (I in e.m.u.)	*m.k.s. units (I in amps)*
$$\frac{2\pi NI}{r} = H_0 \tan \theta$$	$$\frac{NI}{2r} = H_0 \tan \theta$$
$$\therefore I = \left(\frac{rH_0}{2\pi N}\right)\tan \theta$$	$$\therefore I = \left(\frac{2rH_0}{N}\right)\tan \theta$$

A graph of I against $\tan \theta$ is a straight line through the origin. The gradient is the reduction factor. Knowing r and N the value of H_0 may be calculated.

The moving-coil galvanometer

The current enters and leaves at the hairsprings (Fig. 99). In a uniform field of flux density B, the couple acting on a coil of n turns, length l, breadth b and carrying a current I is

$$nBIl \,.\, b \cos \theta = nBIA \cos \theta \quad \text{(Fig. 100)}$$

where A is the area of the coil. If c is the restoring couple in the hairsprings for unit angle of twist then

$$nBIA \cos \theta = c\theta \quad \text{or} \quad I = c\theta/BAn \cos \theta$$

and the scale is not linear.

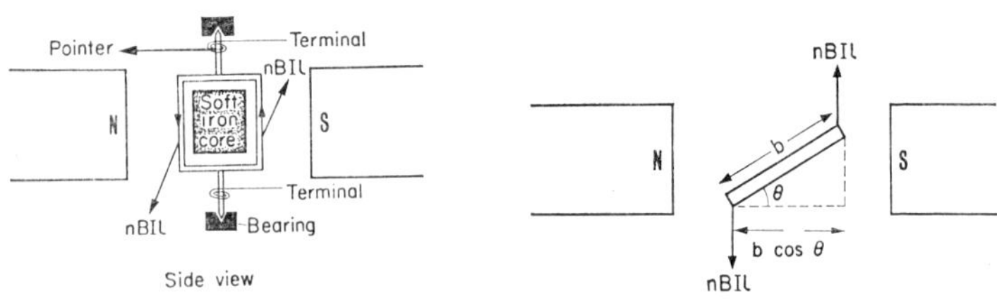

FIG. 99. Moving-coil meter.   FIG. 100

If a soft iron core and curved pole pieces are used (Fig. 101), a radial field is produced and

$$I = c\theta/BnA$$

and the scale is linear.

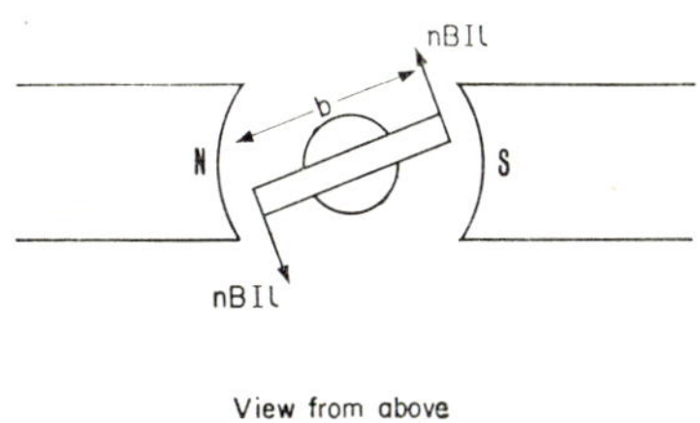

View from above
(radial field)

FIG. 101

The sensitivity is defined as θ/I. For high sensitivity B, n and A must be large and c small. Practical requirements limit the magnitude of n and A. A moving-coil meter may be converted into an ammeter by putting a low resistance shunt in parallel with the meter, and into a voltmeter by putting a high resistance in series.

Electromagnetic Induction

Definitions and laws

157. *Laws of electromagnetic induction.*

(i) When the magnetic flux threading a circuit is made to vary, the induced e.m.f. is proportional to the rate of change of flux linkage (on the m.k.s. system, this law is implicit in the definition of magnetic flux).

$$E = -\frac{d\Phi}{dt}$$

(m.k.s. E in volts, Φ weber-turns. c.g.s. E in abvolts, Φ in maxwell-turns)

(ii) *Lenz's law.* The direction of the induced e.m.f. is such as to oppose the motion or change producing it.

158. The *coefficient of mutual inductance* (*M*) between two circuits is the ratio of the e.m.f. induced in the secondary circuit to the rate of change of current in the primary circuit (usually henries).

$$E = -M\frac{\mathrm{d}I}{\mathrm{d}t}$$

(*I* in amps, *t* in seconds, *E* in volts, *M* in henries)

159. The *coefficient of self-inductance* (*L*) of a circuit (or any part of a circuit) is the ratio of the e.m.f. induced in the circuit to the rate of change of current in the circuit (usually henrys).

$$E = -L\frac{\mathrm{d}I}{\mathrm{d}t}$$

(*I* in amps, *t* in seconds *E* in volts, *L* in henrys)

Verification of the laws of electromagnetic induction

The first law (sometimes known as Faraday's law) may be verified using the apparatus shown in Fig. 102. The copper wire is rotated in the flux of the bar magnet. A graph of volt-meter reading against rate of revolution is a straight line through the origin.

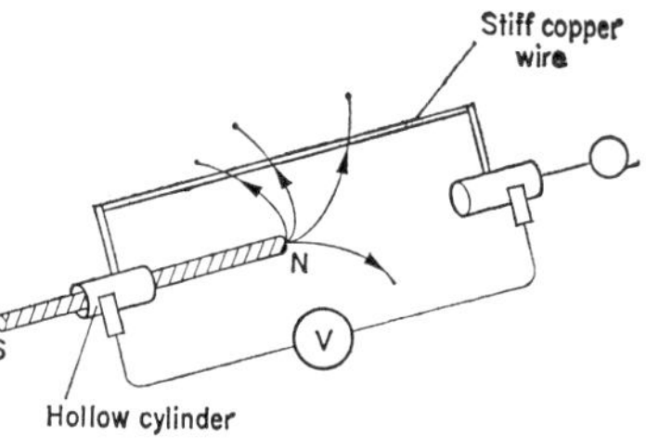

Fig. 102

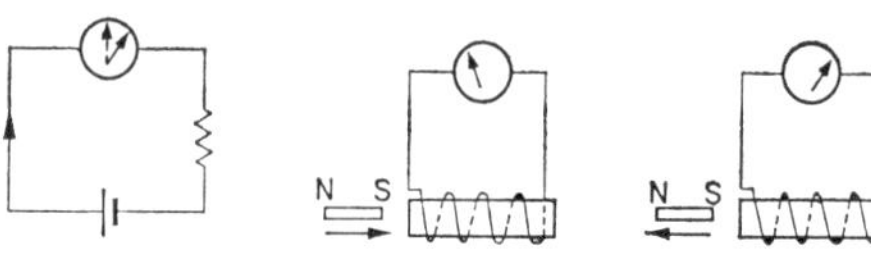

Fig. 103

Lenz's law may be demonstrated as illustrated in Fig. 103. When the current is flowing in the direction shown the galvanomater deflects to the right. The direction of the induced current when the magnet is moved is such that the resulting magnetic field due to the current flowing in the solenoid opposes the motion of the magnet.

Mutual inductance of two concentric solenoids in air

Let the secondary have N_2 turns and be wound on the middle of a primary with N_1 turns per unit length and cross-sectional area A. Then the flux linkage Φ of the secondary due to a current I in the primary is given by

<table>
<tr><td align="center">c.g.s.</td><td align="center">m.k.s.</td></tr>
</table>

$$\Phi = BAN_2$$
$$= (\mu 4\pi N_1 I)AN_2$$
$$\therefore \frac{d\Phi}{dt} = \mu 4\pi N_1 AN_2 \frac{dI}{dt}$$
$$\therefore \underline{M = 4\pi\mu N_1 N_2 A}$$

$$\Phi = BAN_2 = \mu N_1 I A N_2$$
$$\therefore \frac{d\Phi}{dt} = \mu N_1 N_2 A \frac{dI}{dt}$$
$$\therefore \underline{M = \mu N_1 N_2 A}$$

Induced charge due to a flux change

c.g.s.

$$E = -\frac{d\Phi}{dt} \cdot 10^{-8} \text{ volts}$$

Using Ohm's law

$$I = -\frac{1}{R} \cdot \frac{d\Phi}{dt} \cdot 10^{-8} \text{ amps}$$
$$(R \text{ in ohms})$$
$$\therefore \text{ charge } Q = \int I dt$$
$$= -\frac{1}{R} \int_{\Phi_1}^{\Phi_2} d\Phi \cdot 10^{-8} \text{ coulombs}$$
$$\therefore Q = +\frac{\Phi_1 - \Phi_2}{R} \cdot 10^{-8} \text{ coulombs}$$

m.k.s.

$$E = -\frac{d\Phi}{dt} \text{ volts}$$

Using Ohm's law

$$I = -\frac{1}{R} \cdot \frac{d\Phi}{dt} \text{ amps}$$
$$(R \text{ in ohms})$$
$$\therefore \text{ charge } Q = \int I dt$$
$$= -\frac{1}{R} \int_{\Phi_1}^{\Phi_2} d\Phi \text{ coulombs}$$
$$\therefore Q = +\frac{\Phi_1 - \Phi_2}{R} \text{ coulombs}$$

Ballistic galvanometer

This is an instrument whose throw is proportional to the charge which passes through it. It may be calibrated as a flux meter by connecting it in series with the secondary of a standard mutual inductance. The throw θ when known currents are switched off in the primary are recorded. If the resistance of the secondary circuit is R, then

<table>
<tr><td align="center">c.g.s.</td><td align="center">m.k.s.</td></tr>
<tr><td align="center">$Q = \dfrac{\Phi}{R} 10^{-8} = k\theta$</td><td align="center">$Q = \dfrac{\Phi}{R} = k'\theta$</td></tr>
<tr><td align="center">$\therefore\ \Phi = k'R\theta$</td><td align="center">$\therefore\ \Phi = k'R\theta$</td></tr>
</table>

and k' may be determined since Φ is known.

The instrument may now be used to measure flux density by connecting a small search coil in series with the galvanometer. If the area of the coil is A and it has N turns, when it is completely removed from a flux density B,

$$BAN = k'R'\theta$$

for which B may be calculated if the total resistance R' of the galvanometer circuit is known.

Generators

If the coil with n turns (Fig. 104) is rotated at ω radians per second in a uniform field of flux density B then the e.m.f. at any

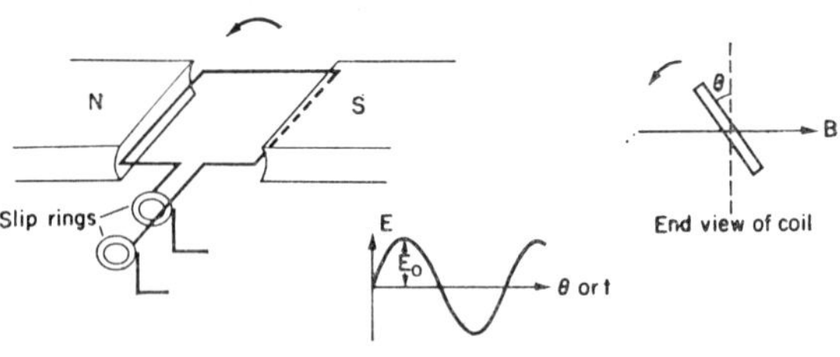

FIG. 104. Alternator.

instant is given by

$$E = \frac{-\mathrm{d}}{\mathrm{d}t}(BAn\cos\theta) = +BAn\sin\theta\,\frac{\mathrm{d}\theta}{\mathrm{d}t}$$

$$\therefore\ E = +BAn\omega\sin\theta = E_0\sin\theta \quad \text{where} \quad E_0 = BAn\omega$$

The d.c. generator is similar in design to an alternator but it has a commutator instead of slip rings.

The d.c. motor

Let V be the p.d. across the armature and I_a the current flowing. If the armature resistance is R_a and E_b is the back e.m.f., then applying Ohm's law, $I_a = (V - E_b)/R_a$. When the motor is first switched on E_b is zero and a starting resistance is used to avoid large currents when starting from rest. Multiplying the above equation by I_a and rearranging we get

$$VI_a = I_a^2 R + I_a E_b$$

VI_a is the power supplied, $I_a^2 R$ is the rate of heat loss and hence, by the conservation of energy, $I_a E_b$ must be the available power. This may also be shown as follows:

$$\text{Torque} = G = BANI_a$$

for a uniform radial field.

$$\therefore\ \text{Rate of doing work} = G\omega = BANI_a \times \omega$$

where ω is the angular velocity.
But $BAN\omega = E_b$

$$\therefore\ \text{Rate of doing work} = I_a E_b$$

Eddy currents

These currents may be demonstrated by oscillating a sheet of copper between the poles of a powerful magnet. The oscillations are heavily damped, but if slots are cut in the sheet, the air gaps act as insulators and reduce the magnitude of the eddy currents. For the same reason, the oscillations of a coil of a suspension type moving-coil galvanometer are heavily damped if the coil is short circuited.

Hysteresis

The fact that B lags behind H (Fig. 105) is known as hysteresis. *Oa* represents the *residual magnetism* or *remanence*, *Ob* the *coercivity*. The area under the curve is the work done *per unit volume* in taking the material completely round a cycle.

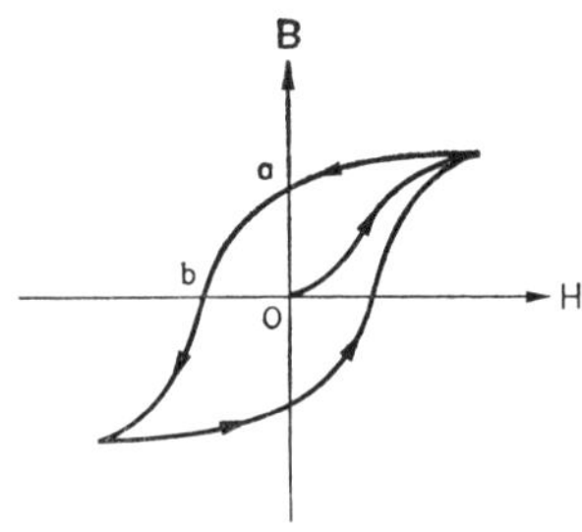

FIG. 105

Soft magnetic materials (e.g. silicon steels) have a thin hysteresis loop and are used for transformer cores and for dynamo-electric machinery. *Hard magnetic materials* (e.g. alnico) are used for permanent magnets; the hysteresis loop is broad and of large area and the materials have high coercivity and high remanence. *Ferrites* are iron compounds which have high resistivities while their magnetic properties are similar to iron. They are used to make high permeability cores for coils as they reduce eddy currents to a minimum.

The transformer

In an ideal transformer the primary has no resistance, all the flux generated in the primary passes through the secondary, and the secondary is on open circuit. Under these conditions if $d\phi/dt$ is the rate of change of flux in the secondary and the primary, and n_s is number of turns on the secondary and n_p the number of turns on the primary

$$\frac{\text{e.m.f. secondary } (E_s)}{\text{e.m.f. primary } (E_p)} = \frac{n_s(d\phi/dt)}{n_p(d\phi/dt)} = \frac{n_s}{n_p}$$

The main losses are (i) flux leakage, (ii) eddy currents (reduced by laminating core), (iii) heating in wires, and (iv) hysteresis loss (reduced by choosing a material with a hysteresis loop of small area).

If no power is lost then $E_s I_s = E_p I_p$ where I_s and I_p are the currents in the secondary and primary respectively. It follows from the above equation that when I_s increases then I_p increases, i.e. when more energy is consumed in the secondary then more energy must be used in the primary.

Electrostatics

Faraday's ice pail experiment shows that a charge always induces an opposite but equal amount of charge on some other conductor (in the case of a so-called isolated charge the induced charge is at "infinity" which means the walls, floor and ceiling of the laboratory).

Experiments with parallel plate capacitors confirm that the induced charge is always equal and opposite to the original charge. We can account for this phenomenon by the concept of electric flux arising from a positive charge and passing through space to terminate on an equal negative charge.

There is no electric field inside a hollow charged conductor. This is a direct consequence of the inverse square law of force between charges. It was demonstrated by Faraday (ice pail experiment) and Cavendish (experiment with two concentric spheres).

(m.k.s. units. Even-numbered pages 98, 100, 102)

Definitions

160. The *capacitance* of a conductor is numerically equal to the charge required to raise its potential by unity (unit: F or C V^{-1}).

161. The *electric field strength* (E) is the negative potential gradient at the point (unit: V m^{-1}). $E = -dV/dx$.

Or, the *electric field strength* at a point is equal to the force per unit charge on a small charge placed at that point (unit: N C^{-1}).

162. *Electric flux* arises from a positive charge and passes through space to the conjugate induced negative charge. The total flux which can arise from a charge Q is equal to Q (unit: C). [This is effectively Gauss's theorem.]

163. If an electric flux Q passes normally through a surface of area A then the *flux density* (D) is defined as Q/A (unit: C m^{-2}).

164. The *electric displacement density* (D) is the charge per unit area which would flow across a conducting sheet placed normal to the field (unit: C m^{-2}). $D = Q/A$.

165. The *permittivity* (ε) of a medium is the ratio of the electric flux density to the electric field strength (unit: F m^{-1}). $\varepsilon = D/E$.

166. The *relative permittivity* (ε_r) is the ratio of the permittivity of the medium to that of free space. The capacitance of a capacitor is increased by this factor when it is filled with the medium instead of there being a vacuum between the plates.

167. A *line of electric force* is a line whose direction at any point represents the direction of the resultant field at the point.

168. The *potential at a point* is the work done per unit charge against the electrical forces in moving a small positive charge from infinity to the point (unit: V).

(**c.g.s. units.** Odd-numbered pages 99, 101, 103)

Definitions

169. The force between two point charges is directly proportional to the product of the charges and inversely proportional to the square of the distance between them; it also depends on the medium.

$$F = \frac{q_1 q_2}{\varepsilon d^2} \text{ dynes}$$

170. One *e.s.u. of charge* (or statcoulomb) repels an identical charge 1 cm away in vacuum with a force of 1 dyn.

171. The *electric field strength* at any point in e.s.u. is numerically equal to the force in dynes which a positive statcoulomb would experience if placed at the point, and its direction is the direction of the force.

172. A *line of electric force* is a line whose direction at any point represents the direction of the resultant field at the point.

173. The *potential at a point* is the work done in ergs against the electrical forces in moving a positive statcoulomb from infinity to the point.

174. The *potential difference* in e.s.u. is the work done in ergs in moving 1 statcoulomb between the points.

175. The *capacitance* of a conductor is numerically equal to the charge required to raise its potential by unity.

176. The capacitance is 1 statfarad if 1 statcoulomb raises its potential by 1 statvolt.

177. The capacitance is 1 *farad* if 1 coulomb raises its potential by 1 volt.

178. The *dielectric constant* (or *permittivity*) is the ratio of the capacitance of a capacitor in that medium to the capacitance of the same capacitor in a vacuum.

(m.k.s. units)

Equivalence of the two definitions of field strength (Fig. 106)

Consider two points A and B whose potentials differ by δV. If F is the force on a charge Q placed between A and B, then the work done in moving a charge Q from B to A is $F(-\delta x) = \delta V \cdot Q$. Therefore

$$E = F/Q = -\delta V/\delta x$$

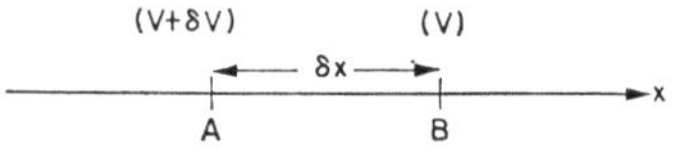

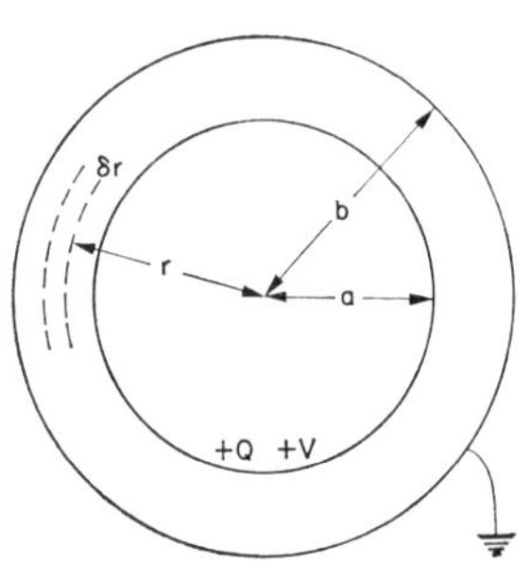

Fig. 106

The capacitance of a parallel plate condenser (Fig. 107)

Let one plate have a charge Q and a potential V. If the plates are a distance d apart and have an area A, then $E = V/d$ and $D = Q/A$.

Substituting in $C = \dfrac{Q}{V} = \dfrac{DA}{Ed} = \dfrac{\varepsilon A}{d}$

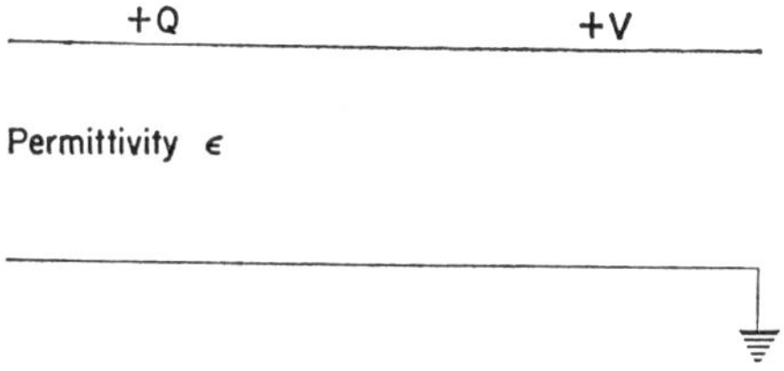

Fig. 107

Capacitance of concentric spheres and concentric cylinders (Fig. 108)

Let the radii be a and b respectively. Let the inner one have a charge Q and potential V. Let the length of the cylinders be l. Then the flux density through a sphere (or cylinder) of radius r ($a < r < b$) is

Fig. 108

(c.g.s. units)

Potential due to a point charge

The work done in moving unit charge a distance δx (Fig. 109) when it is at a distance x from a point charge $+q$ is $q/\varepsilon x^2 \, . -\delta x$. Hence work done in moving unit charge from infinity to a point at a distance r from q is given by

$$\int_{\infty}^{r} -\frac{q}{\varepsilon x^2} \, . \, \mathrm{d}x = \frac{q}{\varepsilon r}$$

Potential at distance r from $+q = \dfrac{q}{\varepsilon r}$

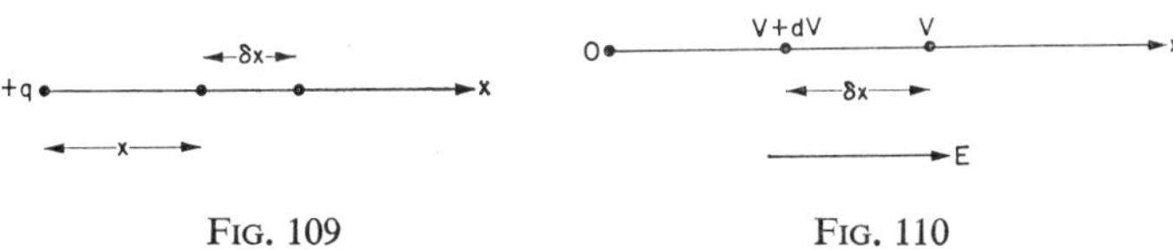

FIG. 109   FIG. 110

The field strength is the negative potential gradient

Consider two points a distance δx apart whose potentials differ by $\mathrm{d}V$ (Fig. 110) in a region where the field strength is E. From the definitions of potential difference and field strength

$$\delta V = E(-\delta x)$$

$$\therefore \; E = -\frac{\mathrm{d}V}{\mathrm{d}x}$$

Gauss's theorem

The total number of lines of induction leaving a charge q is $4\pi q$. This simplified form of Gauss's theorem follows immediately if we take one line of force per unit area to represent unit field strength and εX lines of induction where there are X lines of force.

(m.k.s. units)

<table>
<tr><td align="center">sphere</td><td align="center">cylinder</td></tr>
</table>

$$D = Q/4\pi r^2 \qquad\qquad D = Q/2\pi r l$$

$$\therefore\ E = Q/4\pi r^2 \varepsilon \qquad\qquad \therefore\ E = Q/2\pi r l \varepsilon$$

$$\therefore\ \delta V = \frac{Q}{4\pi r^2 \varepsilon}\, \delta r \qquad\qquad \therefore\ \delta V = \frac{Q}{2\pi r l \varepsilon} \cdot \delta r$$

Integrating and using the fact that $C = Q/V$

$$C = 4\pi\varepsilon \left/ \left(\frac{1}{a} - \frac{1}{b}\right)\right. \qquad\qquad C = 2\pi\varepsilon l / \log_e \frac{b}{a}$$

Coulomb's law

The force between two point charges, Q_1 and Q_2, situated r metres apart is given by

$$F = \frac{Q_1 Q_2}{4\pi\varepsilon r^2}$$

Potential due to a point charge

Using $E = -\mathrm{d}V/\mathrm{d}r$ the potential at a distance a from a point charge of $+Q$ is given by

$$V = -\int_{\infty}^{a} E \cdot \mathrm{d}r = \int_{a}^{\infty} \frac{Q}{\varepsilon \cdot 4\pi r^2} \cdot \mathrm{d}r = \frac{Q}{4\pi a \varepsilon}$$

Capacitance of an isolated sphere

The electric field outside a charged sphere, which is remote from other conductors, is radial, and is indistinguishable from the field which would be produced if all the charge were concentrated at the centre of the sphere.

For a sphere of radius a and charge Q, the field strength E at the surface is (using Coulomb's law or Gauss's theorem) given by $E = Q/4\pi\varepsilon a^2$ and hence $V = Q/4\pi\varepsilon a$. Therefore $C = 4\pi\varepsilon a$.

(c.g.s. units)

Capacitance of a parallel plate capacitor

Let one plate (Fig. 107) have a charge Q and a potential V. If the plates are a distance d apart and have an area A, then

$$E = 4\pi Q/\varepsilon A = V/d$$

By definition $C = Q/V = Q \div 4\pi Q d/\varepsilon A$

$$\therefore \quad C = \frac{\varepsilon A}{4\pi d}$$

Capacitance of an isolated sphere

The lines of force leaving the sphere must be radial otherwise there is a component of force along the surface of the sphere. Therefore the lines of force are the same as if a charge of $+Q$ were at the centre instead of on the sphere. Hence the potential due to the charge Q is Q/r where r is the radius of the sphere. But $C = Q/V = r$ centimetres and is the capacity in e.s.u.

Capacitance of two concentric spheres

Suppose the inner sphere has a charge of $+Q$ on it and is at a potential V (Fig. 108).

Potential of *inner* sphere due to $+Q$ on it $= Q/\varepsilon a$.

Potential of *inner* due to induced charge $-Q$ on outer $= -Q/\varepsilon b$. (Cavendish and later Maxwell demonstrated that there is no field and hence no potential gradient inside a hollow conductor.)

$$\text{Hence} \qquad C = \frac{Q}{V} = \frac{Q}{Q/\varepsilon a - Q/\varepsilon b} = \frac{\varepsilon a b}{(b-a)}$$

Capacitors in parallel

$$C = C_1 + C_2 + C_3.$$

Capacitors in series

$$1/C = 1/C_1 + 1/C_2 + 1/C_3.$$

Energy stored in a charged capacitor

A graph of potential V against charge Q (Fig. 111) is a straight line through the origin.   The energy stored is equal to the work done in charging the condenser. Hence

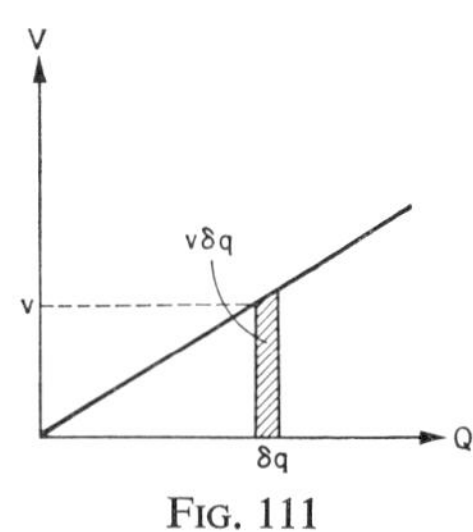

FIG. 111

$$\text{Energy stored} = \text{area under graph} = \int v \, dq = \tfrac{1}{2}QV$$

$$\text{Energy stored} = \tfrac{1}{2}QV = \tfrac{1}{2}CV^2 = \frac{1}{2}\frac{Q^2}{C}$$

Force between two parallel charged plates

The energy stored is equal to $\tfrac{1}{2}QV$ and this energy is lost if the plates come together and discharge. If F is the force between the plates and x their separation then

$$Fx = \tfrac{1}{2}QV \quad \text{and} \quad F = \frac{QV}{2x}$$

Measurement of capacitance

(i) The capacitor of capacitance C_1 is given a charge Q_1 using a known potential V. It is subsequently discharged through a ballistic galvanometer giving a deflection θ_1. Then $C_1V = k\theta_1$. The experiment is repeated using a known capacitor of capacitance C_2 when $C_2V = k\theta_2$, hence $C_1/C_2 = \theta_1/\theta_2$.

(ii) If G (Fig. 112) is a milli-ammeter and the switch is rotated at n revolutions per second, then the current is $I = nCV$ and C may be measured absolutely.

(iii) With an a.c. supply measure the current I passing through the capacitor and the p.d. V across it. Then $V/I = 1/\omega C$ (see page 106)

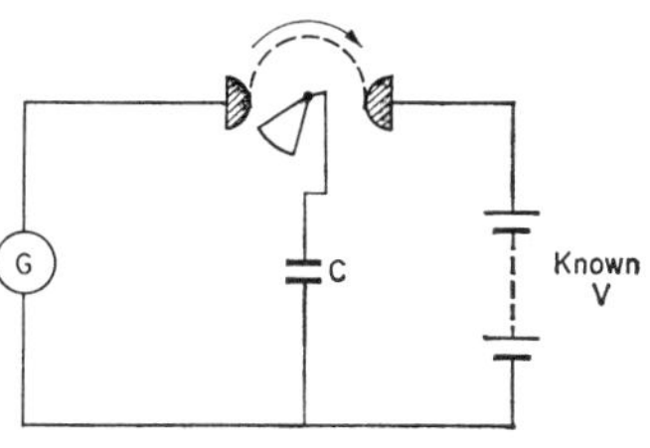

FIG. 112. Measurement of capacitance.

Alternating Current

Measurement of alternating current

(i) A moving-iron meter (coil with low resistance). (ii) A thermocouple meter. (iii) A moving-coil meter in a rectifier bridge circuit (Fig. 113).

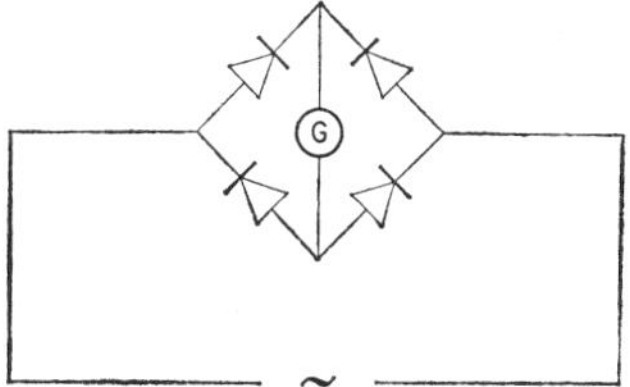

FIG. 113. Bridge circuit for moving-coil meter.

Measurement of alternating voltage

(i) A moving-iron meter (coil with high resistance). (ii) A diode in series with a moving-coil voltmeter. (iii) A moving-coil voltmeter in a bridge rectifier circuit. (iv) An oscilloscope.

Root mean square value of the current (r.m.s. value)

The r.m.s. (or virtual) value of an a.c. is the value of the d.c. which gives the same rate of production of heat as the a.c. This means that it is transferring energy at the same rate. If $i = i_0 \sin\theta$ then $i^2 = i_0^2 \sin^2\theta = i_0^2(1-\cos 2\theta)/2$.

$$\therefore \; i^2 = \frac{i_0^2}{2} - \frac{i_0^2 \cos 2\theta}{2}$$

and since over a complete cycle the average value of $\cos 2\theta$ is zero, the average value of i^2 is $i_0^2/2$. Hence

$$i_{\text{r.m.s.}} = i_0/\sqrt{2}$$

Mean or average current

Average value of a rectified a.c. $= i_{\text{mean}} = \dfrac{2}{\pi} i_0$

$i_{\text{r.m.s.}}/i_{\text{mean}} = 1{\cdot}11$

A.C. circuit containing capacitance only

The charge on the plates at any instant is proportional to the potential difference (V) across the capacitor, and the current flowing is proportional to the rate of change of potential. Therefore at A, B and C (Fig. 114) the current is zero. Notice the current

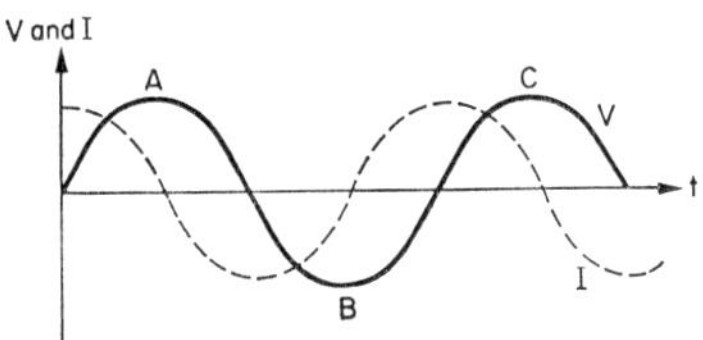

FIG. 114

leads the e.m.f. by $90°$. Since a capacitor must be charged before there is a p.d. across it, the current must lead the p.d. If $V = V_0 \sin \omega t$, then since $Q = CV$,

$$I = \frac{\mathrm{d}Q}{\mathrm{d}t} = C\frac{\mathrm{d}V}{\mathrm{d}t} = C\omega V_0 \cos \omega t = \frac{V_0}{1/\omega C} \sin \left(\omega t + \frac{\pi}{2}\right)$$

i.e. the current leads by $\pi/2$ and the reactance X_c is $1/\omega C$.

A.C. across a pure inductance

Let $i = i_0 \sin \omega t$. If the self-inductance is L, then the applied e.m.f. is $L\,\mathrm{d}i/\mathrm{d}t$. Hence

$$V = L\omega i_0 \cos \omega t = L\omega i_0 \sin (\omega t + \pi/2)$$

i.e. the current lags by $\pi/2$ and the reactance $X_L = L\omega$. Qualitatively, a p.d. must be applied before a current can flow so the p.d. must lead the current.

L, C, R series circuit

The vector diagram is drawn in Fig. 115.

$$V^2 = I^2R^2 + (I\omega L - I/\omega C)^2$$

$$\therefore\ I = \frac{V}{\sqrt{[R^2 + (\omega L - 1/\omega C)^2]}}$$

and

$$\tan\phi = \left(\frac{\omega L - 1/\omega C}{R}\right)$$

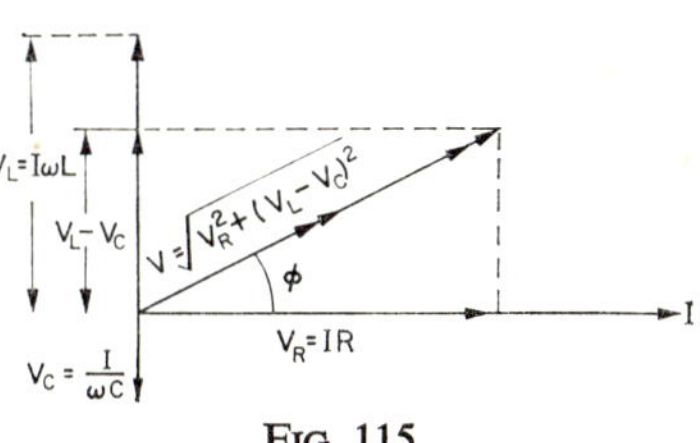

FIG. 115

Valves

Definitions

179. The *anode impedance* (R_a) of a triode is the ratio of a small increase in the anode potential to the increase in the anode current it produces, the grid potential being kept constant.

180. The *amplification factor* (μ) of a triode is the ratio of the increase in anode potential to produce a certain change in the anode current to the increase in grid potential required to produce the same change in anode current.

181. The *mutual conductance* (g_m) of a triode is the ratio of the small change in anode current to the change in grid potential producing it when the anode potential is constant.

182. The *voltage amplification factor* (V.A.F.) or *stage gain* is the ratio of the voltage output to the voltage input.

Thermionic emission

When a metal is heated it emits electrons. The filaments of valves are usually made of tungsten because of its high melting point. Tungsten coated with a mixture of barium and strontium oxide is a very good emitter of electrons. When the anode surrounding the filament is made positive electrons flow from the filament to the anode. Many valves have indirectly heated cathodes.

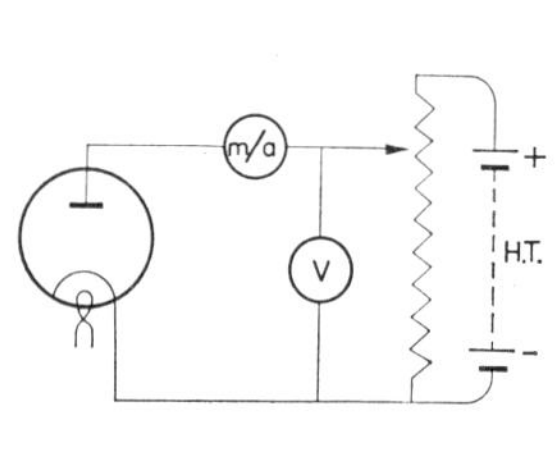

FIG. 116

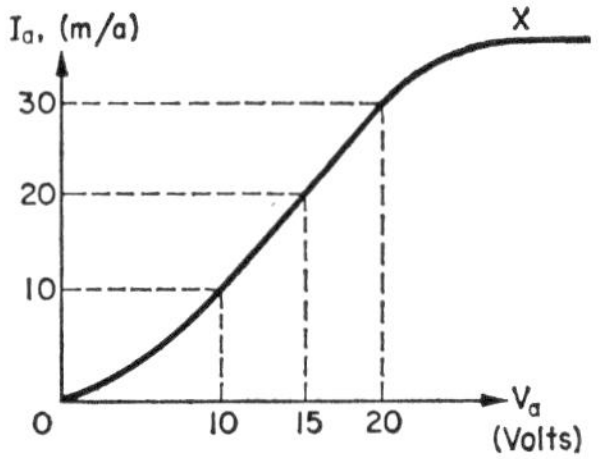

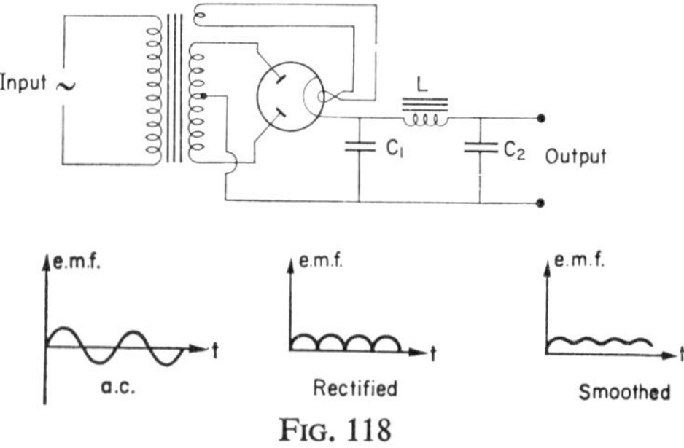

FIG. 117. Characteristics of diode.

Characteristics of a diode

The circuit is shown in Fig. 116 and the resulting graph in Fig. 117. The current is said to be space charge limited up to the point X. The anode impedance of this valve is $10/0{\cdot}01 = 1\ \mathrm{k\Omega}$.

The diode as a rectifier

The double diode provides full wave rectification (Fig. 118).

FIG. 118

The graphs below the circuit show the e.m.f. at the various stages. The reservoir capacitor C_1 charges, and discharges while the e.m.f. is falling thus helping reduce the variations. L and C_2 act as a potential divider for the a.c. ripple voltage thus reducing the ripple further.

Characteristics of a triode

The circuit is shown in Fig. 119, and the curves obtained in Figs. 120 and 121. Using the definitions above and referring to Fig. 121,

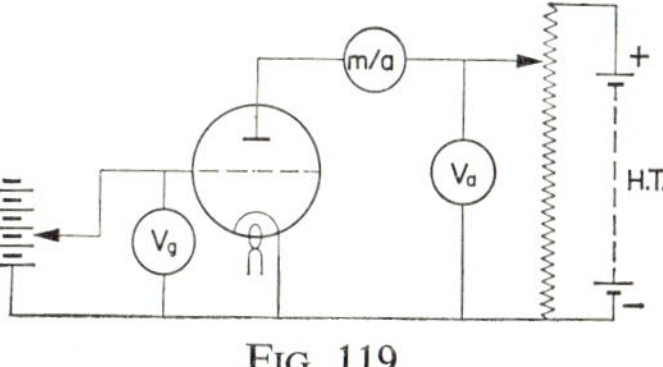

FIG. 119

$$\mu = \left(\frac{\delta V_a}{\delta V_g}\right)_{Ia} = \frac{12}{OC} \qquad R_a = \left(\frac{\delta V_a}{\delta I_a}\right)_{Vg} = \frac{12}{AB}$$

$$g_m = \left(\frac{\delta I_a}{\delta V_g}\right)_{Va} = \frac{AB}{OC}$$

Notice $$\mu = R_a \cdot g_m$$

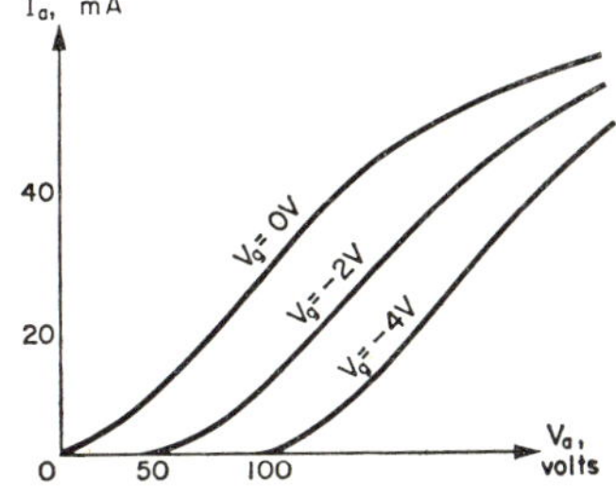

FIG. 120. Characteristics of triode.

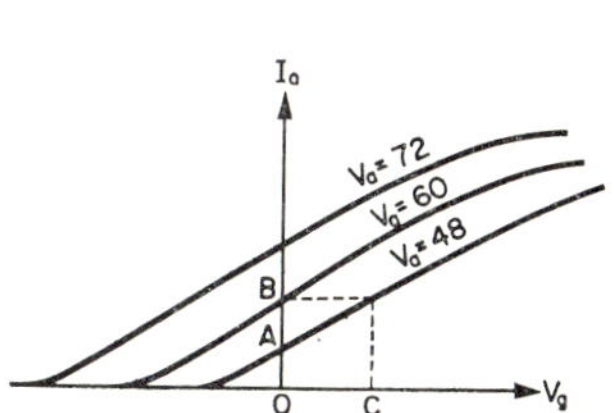

FIG. 121. Characteristics of triode.

The triode as a voltage amplifier (Fig. 122)

A change of V_i on the grid potential will cause the same change in the anode current as a change of μV_i in the potential of the anode circuit. The valve acts as a source of p.d. equal to μV_i with internal resistance R_a. For the alternating component across the load the equivalent circuit is shown in Fig. 123 and the output voltage (V_0) is given by

$$V_0 = \frac{\mu V_i}{(R+R_a)} \cdot R$$

$$\underline{\text{V.A.F.} = V_0/V_i = \mu R/(R+R_a)}$$

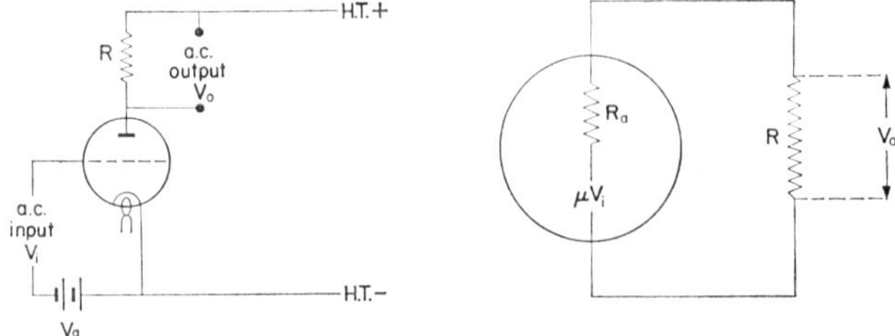

FIG. 122. Triode as voltage amplifier.   FIG. 123

Semiconductors and Transistors

Semiconductors which are *n*-type conduct electricity by electron flow, and those which are *p*-type by positive hole flow. If a strip of *n*-type and a strip of *p*-type are joined in series and a current is passed through them while a magnetic field is applied at right angles to the direction of the current, then measurement of the p.d. developed across opposite faces of each strip (the Hall effect) will determine whether the conduction is mainly due to electrons or to positive holes.

When an *n*-type and *p*-type semiconductor are joined, electrons from the *n*-type flow into the *p*-type and holes from the *p*-type flow into the *n*-type. This migration of charge sets up a potential barrier across the junction. Such an arrangement may be used as a rectifier. If a battery is connected across such a junction with a polarity such that the barrier potential difference is increased, the resistance to current flow across the junction is increased, and the current flow is small, a few microamps. If the polarity of the battery is reversed, it reduces the potential barrier and decreases the effective junction resistance. A current of several milliamps flows for a p.d. of about 1 V.

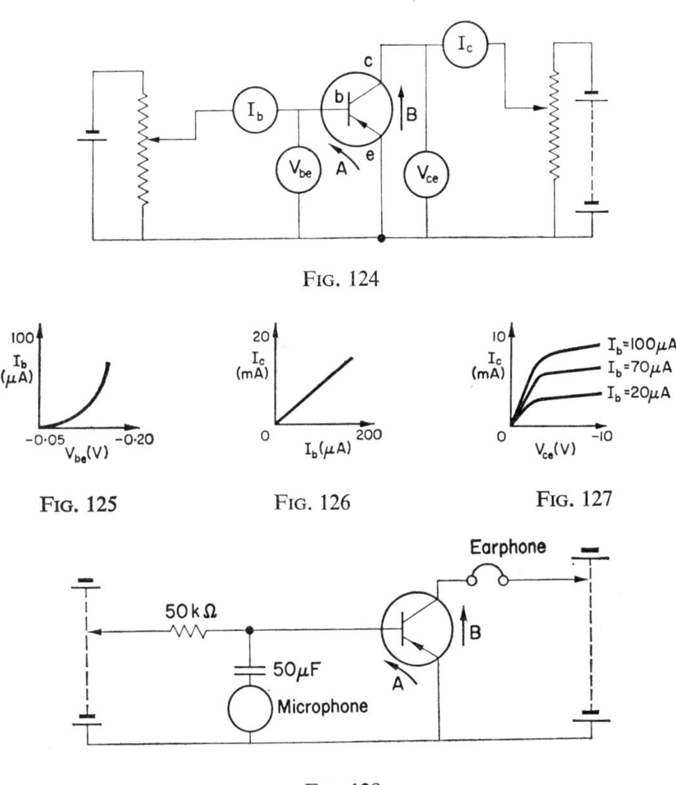

Fig. 124

Fig. 125   Fig. 126   Fig. 127

Fig. 128

Characteristics of a p-n-p transistor in grounded-emitter connection

The circuit is shown in Fig. 124. The *input characteristic* (Fig. 125) and the *transfer characteristic* (Fig. 126) are obtained by keeping the collector–emitter voltage (V_{ce}) constant. The gradient

E

of the transfer characteristic is the *current amplification* factor. The *output characteristic* (Fig. 127) is obtained by keeping I_b constant.

A simple amplifier circuit is shown in Fig. 128. One essential feature of transistors is that a *weak base* current *A* controls a *strong collector* current *B* (Figs. 124 and 128).

Electrons and Photons

Properties of cathode rays

(i) They travel in straight lines (Maltese cross experiment). (ii) They are deflected by electric and magnetic fields, the direction of the deflection indicating that they are negatively charged. (iii) They are negatively charged (Perrin tube experiment). (iv) They cause fluorescence and the liberation of X-rays. (v) They possess considerable energy (a piece of platinum placed in their path will become white hot). (vi) They can penetrate thin metal foil without puncturing it.

Force on an electron moving in a magnetic field

Suppose there are *n* electrons per unit length of beam and that they are moving with a velocity *v*. Charge passing any point per second, that is the current, is *nev*. Substituting in Force = *Bil* we have

$$\text{Force/unit length} = B \,.\, nev$$

$$\therefore \quad \underline{\text{Force on an electron} = Bev}$$

Determination of e/m using the fine beam tube

The electron beam passes through a hole in the anode (Fig. 129). The filament and anode are contained in a glass sphere, the sphere being filled with hydrogen at about 10^{-2} mmHg. At this pressure the positive hydrogen ions formed by the beam of electrons hold the beam together and show its path by means of a weak luminescence caused by the return of the excited hydrogen electrons to the

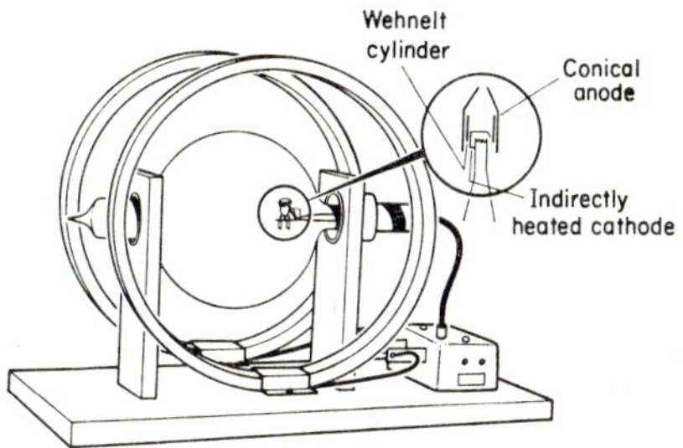

FIG. 129. Determination of e/m.

ground state. The Wehnelt cylinder assists the focusing of the beam. Large Helmoltz coils are arranged on each side of the glass sphere so that the electron beam is in a uniform magnetic field. The electrons moving in the magnetic field are acted on by a force at right angles to their motion and therefore move in a circular path. If the radius of the path is a, the flux density B, and their velocity v then

$$Bev = mv^2/a$$

and B is known from the current flowing in the coils and their dimensions. a may be measured with a ruler. If the electrons are accelerated through a p.d. of V volts between the anode and cathode then

$$\tfrac{1}{2}mv^2 = eV$$

$$\therefore\ e/m = 2V/B^2a^2$$

$\begin{cases} \text{m.k.s. } e \text{ coulombs, } m \text{ kilograms, } V \text{ volts, } B \text{ webers per square metre, } a \text{ metres} \\ \text{c.g.s. } e \text{ e.m.u., } m \text{ grams, } V \text{ e.m.u., } B \text{ gauss, } a \text{ centimetres} \end{cases}$

Millikan's determination of e (Fig. 130)

The oil drops become charged by friction as they pass through the atomizer. The velocity v_1 is first determined as a drop is observed through a microscope while it is falling freely under gravity. If $m'g$ is the apparent weight, η the coefficient of viscosity of air, a the radius of the drop, ρ the density of oil and σ the density of air, we have, applying Stokes's law

$$m'g = 6\pi\eta a v_1 = \tfrac{4}{3}\pi a^3(\rho - \sigma)g$$

and hence a may be found and shown to be constant.

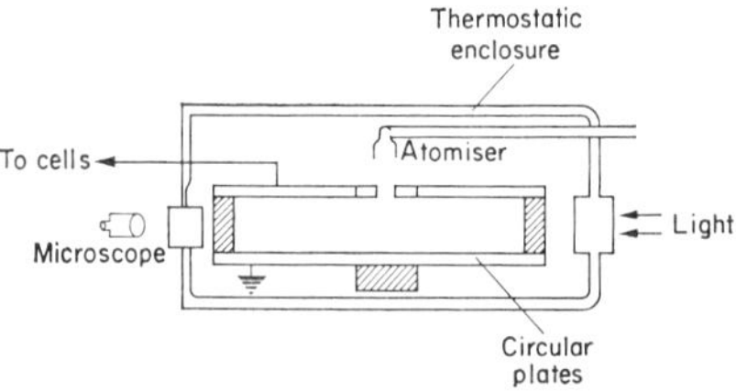

FIG. 130. Millikan's apparatus for e.

The upward velocity v_2 is measured when a potential difference V is applied across the plates. If d is the separation of the plates, then the field strength is $E = V/d$. We have

$$Ene - m'g = 6\pi\eta a v_2$$

where ne is the charge on the drop.

Adding the two equations

$$ne = \frac{6\pi\eta a(v_1 + v_2)}{E}$$

and ne is always a simple multiple of $4\cdot8 \times 10^{-10}$ e.s.u. ($1\cdot6 \times 10^{-19}$ coulomb).

$\begin{cases} \text{m.k.s. } e \text{ coulombs, } \eta \text{ Newton sec m}^{-2}, a \text{ metre, } v_1, v_2 \text{ m/sec, } E \text{ volts/m} \\ \text{c.g.s. } e \text{ statcoulombs, } \eta \text{ dyne sec cm}^{-2}, a \text{ centimetre, } v_1, v_2 \text{ cm/sec, } E \text{ statvolt/cm} \end{cases}$

The cathode ray tube (Fig. 131)

Electrons are emitted from a heated filament F. The Wehnelt cylinder (or grid) is at a negative potential to collimate the beam; it controls the brightness. The anodes (A_1, A_2 and A_3) focus the beam before they pass the X plates (in a vertical plane) and the Y plates (in a horizontal plane).

Uses. (i) A voltmeter. It can be used for d.c. or the amplitude of any a.c. voltage (it is a voltmeter of infinite resistance and takes no current). (ii) The study of wave forms. A *time-base* is connected across the X plates. This provides a p.d. which increases linearly so that the beam moves across the screen, and then drops to zero when the beam reaches the other side of the screen. The wave form to be studied is fed to the Y plates. (iii) Television. (iv) The measuring of short time intervals, e.g. radar.

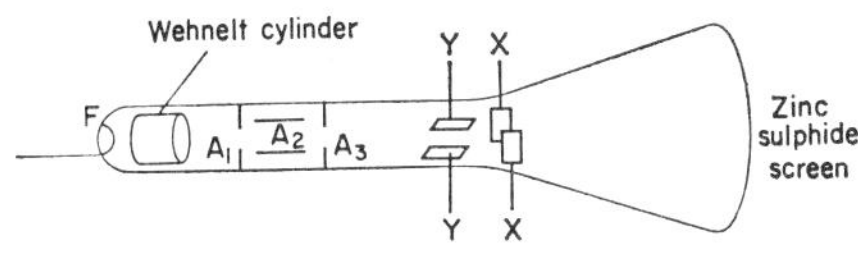

FIG. 131. Cathode ray tube.

Definition

183. An *electron volt* (eV) is the energy acquired by an electron when it moves through a p.d. of 1 volt.

The photoelectric effect

When light shines on sodium, the sodium emits electrons. In fact all metals exhibit this effect if light of short enough wavelength is used. Most metals need ultraviolet light. The following rules apply to photoelectric emission. (i) The number of electrons emitted is proportional to the intensity of the incident light. (ii) The maximum energy E of the electrons increases with frequency v of the incident light and $E = hv$. Below the threshold frequency no electrons are emitted. The minimum energy needed by an electron to escape from the surface of a metal is known as the *work function* of the metal.

The above observations may be explained by the quantum theory. Suppose the energy needed to free an electron is E_0, then the energy of the emitted electron is $hv - E_0$, where h is Planck's constant and v the frequency of the light. This predicts, as expected, that the velocity of the electrons depends on the frequency of the light. Figure 132 shows in a diagrammatic way how the relationship may be tested. A repulsive electric field just sufficient to stop the electrons is applied. If this potential is V volts and the charge on an electron is e coulombs, then the energy of emission is Ve joules. Hence $Ve = hv - E_0$. A graph of V against v is a straight line as predicted by the quantum theory. The gradient of the graph is h/e.

The change of energy (hv) on the collision of a photon with an electron has been observed (the Compton effect).

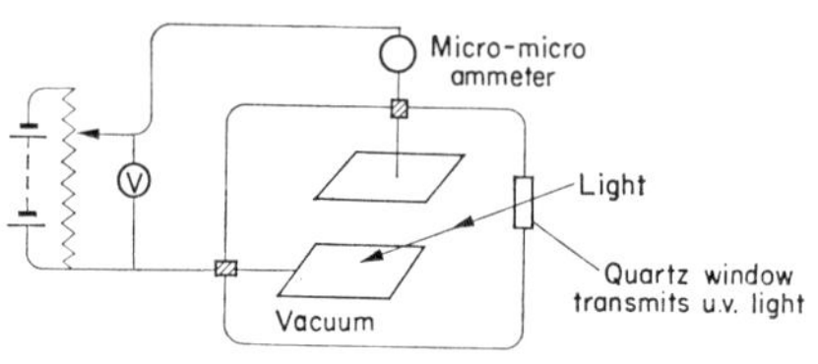

FIG. 132

Waves or particles?

The quantum theory explains (i) the distribution of energy in the spectrum of radiation emitted by a hot body, (ii) the variation of the specific heats of solids with temperature, (iii) the photoelectric effect, (iv) the production of line and band spectra by excited atoms and molecules. The wave and particle nature of light have now been combined in the equations of quantum mechanics. We may think of interference as caused by photons guided to the bright patches by waves.

Wave–particle duality is universal and the diffraction of electrons has been observed.

CHAPTER 6

Nuclear Physics

Definitions

184. The *atomic weight* of an element is 12 times the ratio of the weight of one atom of it to the weight of one atom of carbon 12.

185. The *mass number* of an element is the number of protons and neutrons in the nucleus of one atom of it.

186. The *atomic number* of an element is the number of protons in the nucleus of one atom of it.

187. *Isotopes* of an element are atoms with the same atomic number but with different atomic weights.

188. *Isobars* are atoms with the same atomic weights but with different atomic numbers.

189. *Radioactivity* is the disintegration of atomic nuclei with the emission of ionizing particles or rays.

190. *Law of radioactive decay.* The number of disintegrations per unit time in a sample of a radioactive substance is directly proportional to the number of atoms of the substance present, and is independent of the temperature and pressure.

191. The *half-life* for a radioactive disintegration process is the time taken for the number of disintegrations per unit time to fall to half its initial value.

192. A *curie* is the amount of material in which $3 \cdot 7 \times 10^{10}$ atoms disintegrate per second.

Radioactivity

Radioactivity may be detected by a spinthariscope, a cloud chamber, a solid state detector, a Geiger–Müller tube, or an ionization chamber connected to either a Wulf electroscope or d.c. amplifier and current meter. When a Wulf electrometer (Fig. 133) is used the rate of discharge of the leaf against the counter electrode is a measure of the ionization current.

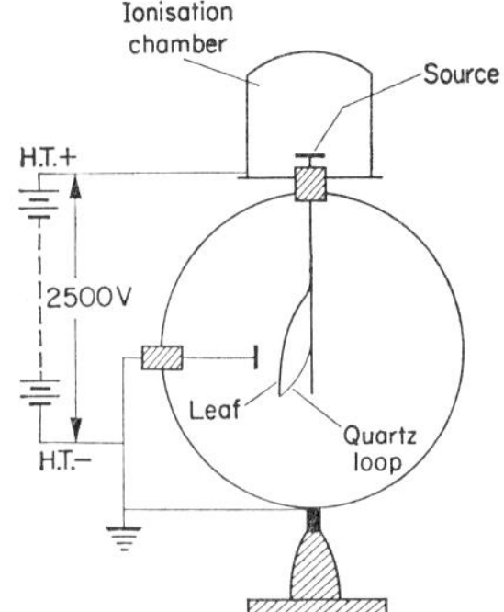

FIG. 133. Wulf electroscope.

Types of radiation

α-particles are emitted mainly by elements of atomic number greater than 82. β-emission is characteristic of nuclei which have an excessive proportion of neutrons; the effect of the process is the conversion of one of the neutrons in the nucleus to a proton. Positive β-rays are positrons. They are emitted by nuclides which have an excessive proportion of protons. The effect of the process is the conversion of a proton in the nucleus to a neutron. γ-rays are emitted when an excited nucleus returns to the ground state. α- and β-emitters often emit γ-rays as well, since the emissions often produce nuclei in excited states.

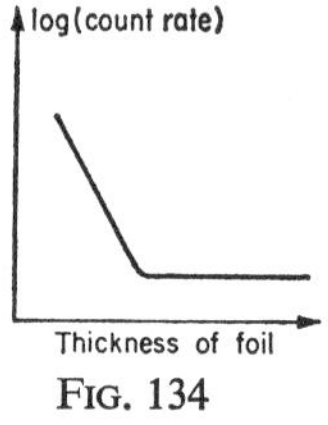

FIG. 134

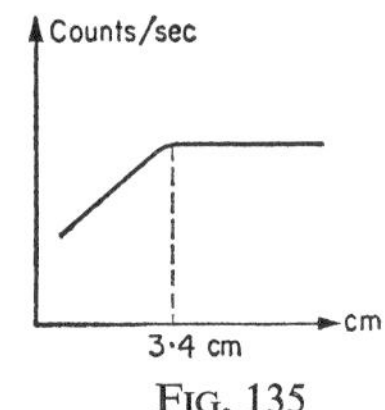

FIG. 135

Evidence for α- and β-particles

Add foils of thin aluminium one at a time on top of the source of radium which is in an ionization chamber or in front of a solid state detector. The graph obtained is shown in Fig. 134. The break in the graph shows where the α-particles are cut off.

Range of α-particles in air

The outer cylinder is replaced by a telescopic chamber. The count rate is plotted against the distance of the top of the chamber from the source (Fig. 135); alternatively, the distance of the source from a solid state detector may be varied. The range of α-particles from a radium source in air is about 3·4 cm.

Geiger–Müller tube

The Geiger–Müller tube may be connected either to a *scalar* (counts particles entering the tube) or to a *ratemeter* (average rate at which particles enter is shown in a dial).

Evidence for γ-radiation

Thin lead foils are placed over the source and the count rate is determined using a Geiger–Müller tube. The break in the graph (Fig. 136) indicates when all the β-particles have been absorbed.

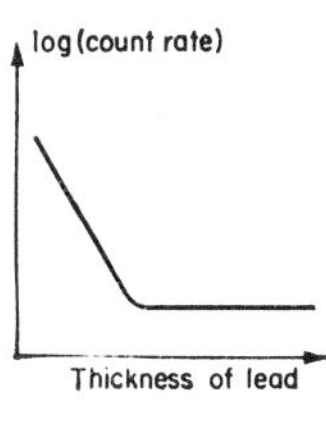

FIG. 136

Properties of α-radiation

Easily absorbed (a few centimetres of air at atmospheric pressure). Deflected by electric and magnetic fields. Helium ions with a definite velocity. Figure 137 illustrates Rutherford and Royd's experiment to show that α-radiation is helium ions. The radon gas is left for some days and then the reservoir is raised to compress any gas which might be present. A helium spectra is observed in the capilliary discharge tube.

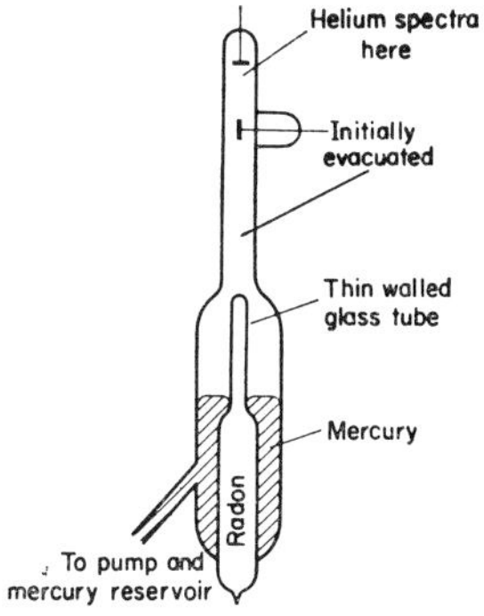

FIG. 137. Rutherford and Royd's experiment.

Properties of β-radiation

Electrons. Wide range of velocities. Fairly easily absorbed (a few millimetres of aluminium foil will absorb strong β-radiation). Easily deflected by magnetic and electric fields.

Properties of γ-radiation

Electromagnetic waves of shorter wavelength than X-rays. Unaffected by electric and magnetic fields. Radiation from strong γ sources is very penetrating.

Law of radioactive decay (definition 190)

This may be easily verified by using $^{220}_{86}$Rn (a decay product of $^{232}_{90}$Th) which has a half-life of $54\frac{1}{2}$ sec; plot the logarithm of the activity against the time.

If there are N atoms present at a time t and λ is the fraction of these that disintegrate in unit time then $\delta N = -\lambda N \delta t$. Integrating $\log_e (N/N_0) = -\lambda t$ or $N = N_0 e^{-\lambda t}$; λ is called the *decay constant*. From the definition of the half-life T,

$$\log_e 0{\cdot}5 = -\lambda T \quad \text{or} \quad \lambda T = \log_e 2 = 2{\cdot}303 \log_{10} 2 = 0{\cdot}693.$$

That is

$$T = \frac{0{\cdot}693}{\lambda}$$

Also $\qquad\qquad N_0/2 = N_0 e^{-\lambda T} \qquad \therefore\ e^{-\lambda} = (\tfrac{1}{2})^{1/T}$

substituting in $N = N_0 e^{-\lambda t}$ we get

$$N = N_0(\tfrac{1}{2})^{t/T}$$

Half-life of radon-220

Squeeze some radon-220 into an ionization chamber connected to a d.c. amplifier and a current meter. The time for the count rate to fall by half may be determined. A graph of $\log_{10}$ (count rate) against time is linear. The time for the logarithm to change by $0{\cdot}301$ is the half-life.

Half-life of a source in a decay series

When radioactive equilibrium is reached, the rate of formation of a short-lived isotope is equal to the rate of decay, hence $\lambda_1 N_1 = \lambda_2 N_2$, and $N \propto T$. If A is the atomic mass, then the mass of an element present $m \propto AT$, hence

$$m_1/m_2 = A_1 T_1/A_2 T_2$$

and the half-life of one isotope may be determined from the relative abundance if the half-life of the other is known.

Uses. (i) Medical: radioactive iodine for the thyroid gland and the treatment of cancer. (ii) Agriculture: the study of the assimilation of different chemicals by plants. (iii) Industry: automatic control of the thickness of products and the measuring of piston wear.

Rules of radioactive decay

(i) When an element disintegrates by the emission of an α-particle it turns into an element with chemical properties similar to those of an element two places earlier in the periodic table. $^{238}_{92}\text{U} = ^{234}_{90}\text{Th} + _2\text{He}$. (ii) When an element disintegrates by the emission of a β-particle it turns into an element with properties similar to those of an element one place later in the periodic table. $^{239}_{92}\text{U} = ^{239}_{93}\text{Np} + _{-1}^{0}\text{e}$.

Nuclear Fission

Certain types of heavy nucleus ($^{235}_{92}\text{U}$ in particular) decay by breaking into two approximately equal halves. Much energy and a certain number of free neutrons are given off. The neutrons can trigger the splitting of further nuclei and hence a chain reaction may take place. Uranium fission may be represented by the equation

$$^{235}_{92}\text{U} + _0^1 n = ^{236}_{92}\text{U} = ^{144}_{56}\text{Ba} + ^{90}_{36}\text{Kr} + 2 \,.\, _0^1 n + \text{heat energy}$$

Uses. (i) Uncontrolled chain reaction: atomic bombs. (ii) Controlled chain reaction: nuclear reactors.

Nuclear Fusion

Under conditions of extreme heat and pressure light nuclei may be made to combine to form heavier nuclei. Some of these reactions are highly exothermic, for example

$$\mathrm{^2_1H + {}^2_1H = {}^4_2He} + 5 \cdot 7 \times 10^{11} \ \mathrm{J \ g^{-1}}$$

The energy given off can be used to maintain the reaction.

Uses. (i) Uncontrolled: hydrogen bomb and production of energy of Sun and stars. (ii) Controlled: this has not yet been achieved (1968) but no doubt when it is it will be used to generate electric power.

X-rays

Production of X-rays

They are produced when fast moving electrons are rapidly decelerated, e.g. by striking a target. The energy of the electron is converted into electromagnetic waves of very short wavelength (about 10^{-8} cm). A beam of electrons is emitted from a hot filament in a vacuum tube. The electrons are accelerated towards a piece of tungsten or molybdenum which is at a high potential. When the electrons strike the metal X-rays are emitted.

Properties of X-rays

(i) Very penetrating. (ii) They can be reflected, refracted, diffracted and polarized. (iii) They ionize gases. (iv) They affect photographic plates. (v) They cause fluorescent substances such as zinc sulphide to glow.

X-ray spectra

They consist of intense lines characteristic of the metal (due to energy changes resulting from deep penetration of bombarding electrons to K and L shells of the atom) superimposed on a background of continuous or "white" radiation of relatively weak intensity. For low voltages across the tube no characteristic lines appear. Moseley's law states

$$v = a(Z - b)^2$$

where v = frequency of characteristic emission lines, Z = atomic number, and a and b are constants.

CHAPTER 7

Absolute Measurements

Current balance or dynamometer (Fig. 138)

The current is passed through C_1 alone and the pointer adjusted to be horizontal. The current I is then passed through both coils in series and the pointer again adjusted to be horizontal by moving the rider.

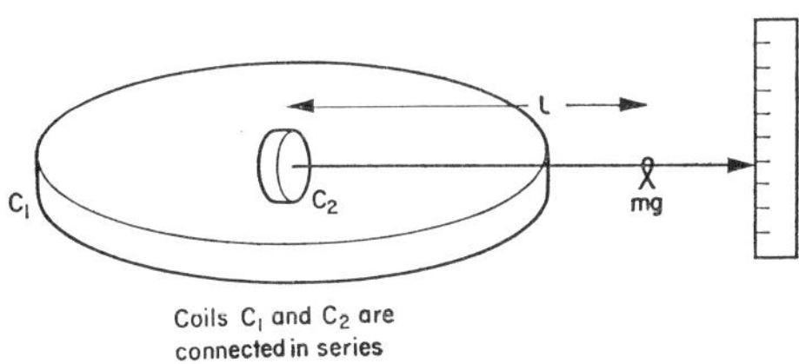

FIG. 138. Current balance.

If R_1 is the radius of C_1 (N_1 turns) and R_2 is the radius of C_2 (N_2 turns), then

c.g.s.	m.k.s.
Field at centre of C_1	Flux density at centre of C_1
$$= \frac{2\pi N_1 I}{R_1}$$	$$= \frac{\mu . N_1 I}{2R_1}$$
Couple acting on C_2	Couple acting on C_2
$$= \frac{2\pi N_1 I}{R_1} . A I N_2$$	$$= \frac{\mu N_1 I}{2R_1} . A I N_2$$

where A is the cross-sectional area of C_2.

$$\therefore\ mgl = \frac{2\pi N_1 I}{R_1} \cdot \pi R_2^2 \cdot IN_2$$

$$\therefore\ I = \sqrt{\left(\frac{mgl \cdot R_1}{2\pi^2 R_2^2 N_1 N_2}\right)}$$

where A is the cross-sectional area of C_2.

$$\therefore\ mgl = \frac{\mu N_1 I}{2R_1} \cdot \pi R_2^2 \cdot IN_2$$

$$\therefore\ I = \sqrt{\left(\frac{2mglR_1}{\mu\pi R_2^2 N_1 N_2}\right)}$$

The absolute determination of the ohm (Fig. 139)

The p.d. between the rim and axle of a disc which is rotating in the magnetic field of a solenoid is balanced against the p.d. across a resistor. If the disc rotates at f revolutions per second, and a is its radius, N the number of turns per unit length on the solenoid and I the current flowing. Then

c.g.s.

Induced e.m.f.

$$= 4\pi NI \cdot \pi a^2 f = IR$$

$$\therefore\ R = 4\pi a^2 Nf$$

m.k.s.

Induced e.m.f.

$$= \mu NI \cdot \pi a^2 f = IR$$

$$\therefore\ R = \mu\pi a^2 Nf$$

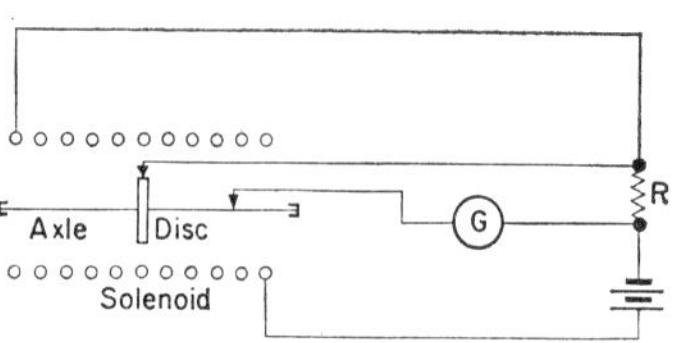

FIG. 139. Absolute determination of the ohm.

The above calculation is oversimplified and in practice allowance must be made for the lack of uniformity of field and for the Earth's field. Thermoelectric e.m.f.'s are troublesome.

The attracted disc electrometer (Fig. 140)

The force of attraction between the plates is balanced by the force mg. If V is the potential of the top plate with a charge Q on it then:

$$\tfrac{1}{2}Q\frac{V}{d} = mg. \qquad \text{Also} \quad Q = CV$$

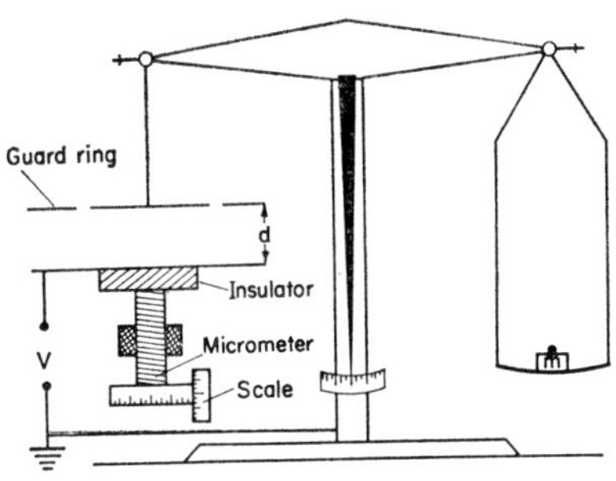

FIG. 140. Absolute determination of p.d.

c.g.s.	m.k.s.

$$C = A/4\pi d \quad \therefore \quad Q = VA/4\pi d \qquad\qquad C = \varepsilon A/d \quad \therefore \quad Q = V\varepsilon A/d$$

$$\therefore \frac{1}{2}\frac{V \cdot A}{4\pi d} \cdot \frac{V}{d} = mg \qquad\qquad \therefore \frac{1}{2}\frac{V\varepsilon A}{d} \cdot \frac{V}{d} = mg$$

$$\therefore V = \sqrt{\left(\frac{8\pi d^2}{A} \cdot mg\right)} \qquad\qquad \therefore V = \sqrt{\left(\frac{2d^2}{\varepsilon A} \cdot mg\right)}$$

The amp and the ohm are the two absolute determinations that are actually made. The attracted disc electrometer is not sufficiently accurate because (i) the balance is unstable for small values of d, and (ii) the force is small and difficult to measure accurately.

CHAPTER 8

Additional Topics in Some Syllabuses

e/m for Electrons by Electric and Magnetic Deflection (Fig. 141)

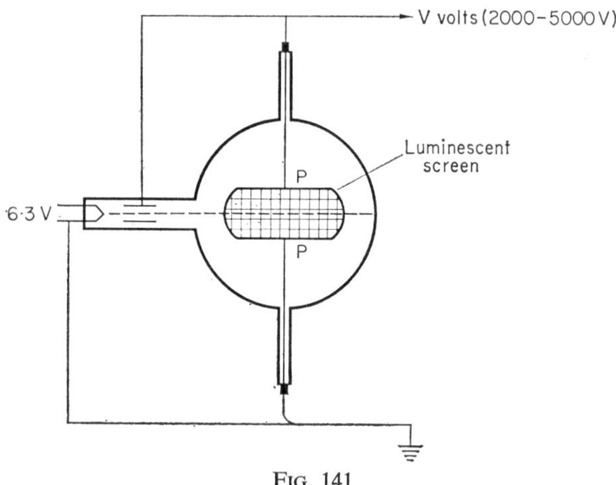

FIG. 141

A beam of cathode rays passes obliquely over a screen, one side of which is coated so that it is luminescent and over-printed with a centimetre scale. An electric field E is applied between PP. A magnetic field of flux density B may be applied perpendicular to the paper. When the beam is undeflected, the force on the electron due to the electric field E is balanced by the force due to the magnetic flux density B, and if v is the velocity of the electrons

$$Ee = Bev \quad \therefore \quad v = \frac{E}{B}$$

If the electron is accelerated through a p.d. of V then

$$\tfrac{1}{2}mv^2 = eV \tag{17}$$

and

$$\frac{e}{m} = \frac{v^2}{2V} = \frac{E^2}{2B^2V}$$

Alternatively if r is the radius of the path when in the magnetic field

$$Bev = \frac{mv^2}{r} \tag{18}$$

Combining this with eqn. (17)

$$\frac{e}{m} = \frac{2V}{B^2r^2}$$

This method is the basis of e/m measurements for fundamental particles.

e/M for Positive Ions

Thomson method. Thomson passed a beam of positive ions through parallel electric and magnetic fields. If the ions are moving with velocity v parallel to the x-axis (Fig. 142) and the

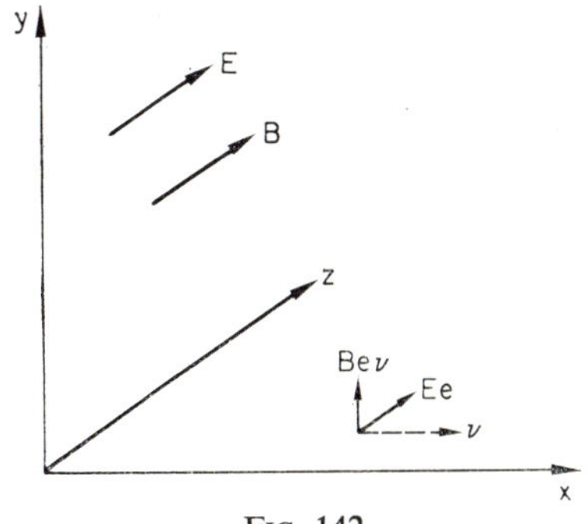

FIG. 142

electric and magnetic fields (E and B) are as shown, then using eqn. (2) on page 4 we have:

Displacement of ion in z direction $= \dfrac{1}{2}\dfrac{Ee}{M}\left(\dfrac{x}{v}\right)^2.$

Displacement of ion in y direction $= \dfrac{1}{2}\dfrac{Bev}{M}\left(\dfrac{x}{v}\right)^2.$

(for small deflections)

Eliminating v we have

$$y^2 = \frac{x^2 B^2}{2E}\cdot\frac{e}{M}\cdot z$$

$$\therefore\ y^2 = \text{constant} \times \frac{e}{M} \times z$$

A parabola is therefore formed on the screen for ions of particular e/M, and from measurements of the co-ordinates e/M may be calculated.

Bainbridge method (Fig. 143). S_1, S_2, S_3 are slits. An electric field (E) exists between the plates P_1 and P_2. A perpendicular

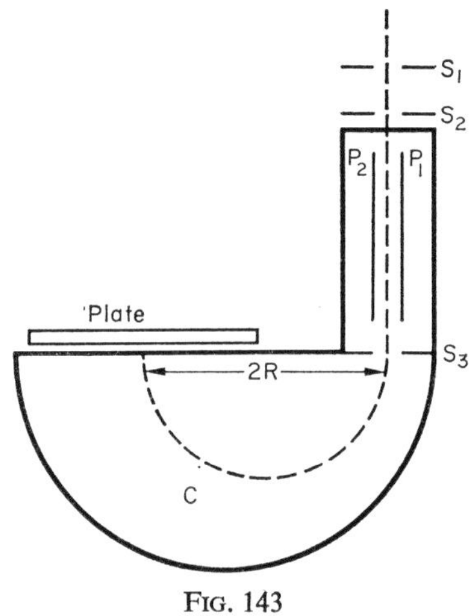

FIG. 143

magnetic field (B) extends over the same region. Only those ions with velocity $v = E/B$ will enter the deflection chamber C. A strong magnetic field B' extends over the chamber which bends the ions into a semicircular path. If R is the radius of this path, we have (eqn. (18) page 127)

$$\frac{e}{M} = \frac{v}{B'R} = \frac{E}{B'BR}$$

The Triode as an Oscillator

The resonant frequency of the LC circuit (Fig. 144) is given by $f = 1/2\pi\sqrt{(LC.)}$ The coils L and L' are inductively coupled so that the alternating current in the LC circuit provides an alternating p.d. to the grid. The valve therefore supplies an alternating current to the resonating circuit so that losses due to resistance are made up.

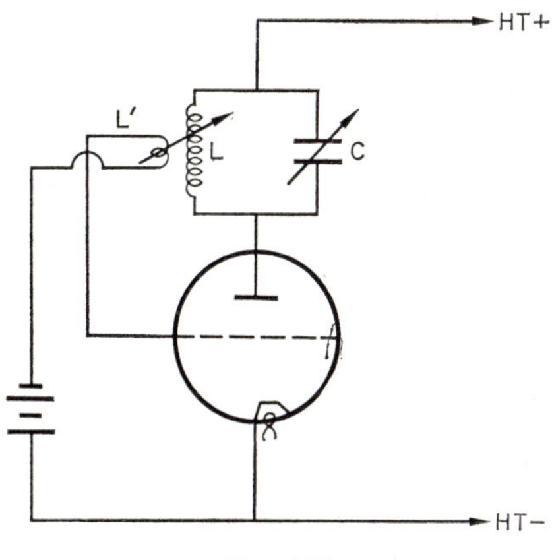

FIG. 144

Atomic Structure

α-particle scattering experiments led to the abandonment of the Thomson "plum pudding" model of the atom, and the establishing of a model of the atom as a positive nucleus surrounded by orbital electrons.

A modern "picture" of the atom, the electron cloud model, is derived from Schrödinger's wave equation. The density of the cloud represents the probability of finding an electron at a given position.

The energy required to raise the atom from its ground state to an excited state is called the *excitation energy*. If this energy is eV electron volts, then V is the *excitation potential*. The corresponding energy and potential needed to remove an electron which is in the ground state from an atom are called the *ionization energy* and the *ionization potential*.

Faraday's Laws of Electrolysis

1. The mass of a substance liberated during electrolysis is proportional to the quantity of electricity which has passed.

2. The quantities of electricity required to liberate 1 g-atom of different elements bear a simple relationship to one another.

Radiation

Black body and Stefan's law

A perfectly *black body* is one which absorbs completely every wavelength incident on it.

Stefan's law. The total radiation (E) from a given body is directly proportional to the fourth power of the absolute temperature of that body.

For a body at temperature T_1 °K, in surroundings at temperature T_2 °K, then

$$\text{net rate of energy loss} = \sigma A(T_1{}^4 - T_2{}^4) \quad \text{J s}^{-1} \text{ deg K}^{-4}$$

where σ is Stefan's constant and A the surface area. It is only strictly true for a small black body in a large enclosure.

Distribution of energy in the spectrum of a black body

This may be investigated using a spectrometer with mirrors (glass absorbs radiation) and a suitable grating. The curves obtained are shown in Fig. 145.

The characteristics are:

(i) Areas between curves and λ-axis rise rapidly ($\propto T^4$).

(ii) For every temperature there is a maximum value of e_λ. As T rises this value moves towards shorter wavelengths.

(iii) $\lambda_{max} \propto 1/T$ (Wien's law).

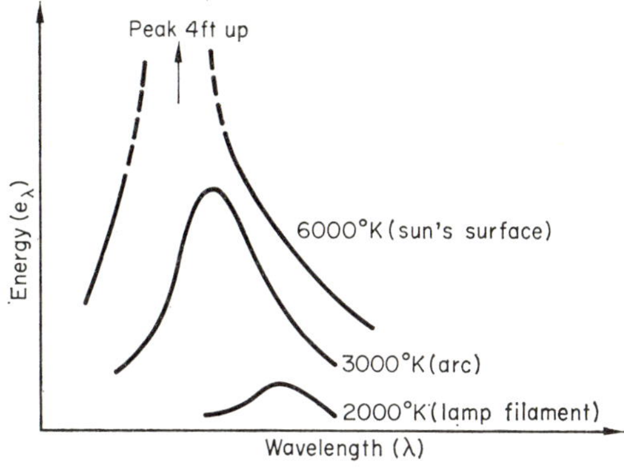

FIG. 145

Total radiation pyrometer (Fig. 146)

The mirror M is moved until an image of the high temperature source is obtained on D. Provided the image covers the whole aperture in D, the radiation passing through this hole is independent of the distance from the source. The instrument is calibrated so that the e.m.f. produced by the thermocouple indicates the temperature of the source.

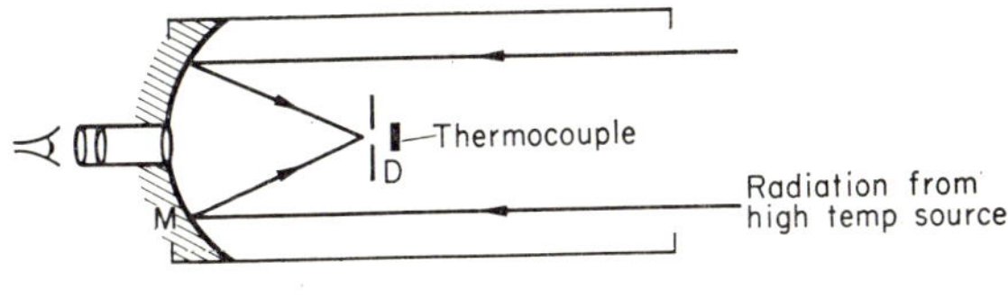

FIG. 146

Optical pyrometer (Fig. 147)

The image of the hot source is focused on the filament of an electric lamp. This image and the filament are viewed through a red filter. The current through the lamp is adjusted until the filament is indistinguishable from the image. The instrument is calibrated so that the current through the lamp measures the temperature of the source.

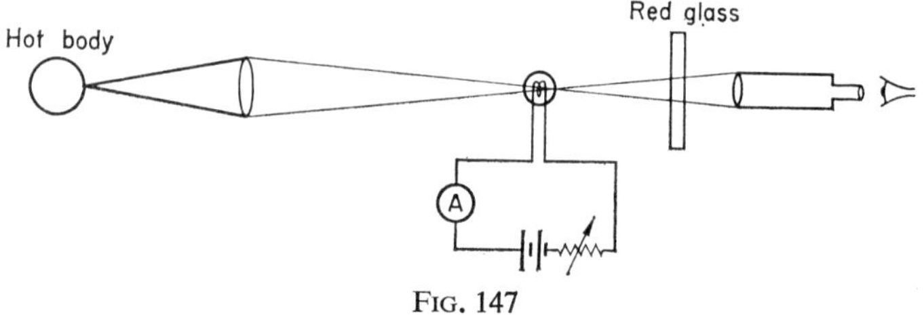

Fig. 147

Liquefaction of Gases

There are two main methods.

(i) In the Linde process (Fig. 148) air is compressed to a high pressure (100–200 atmospheres). During this operation it heats

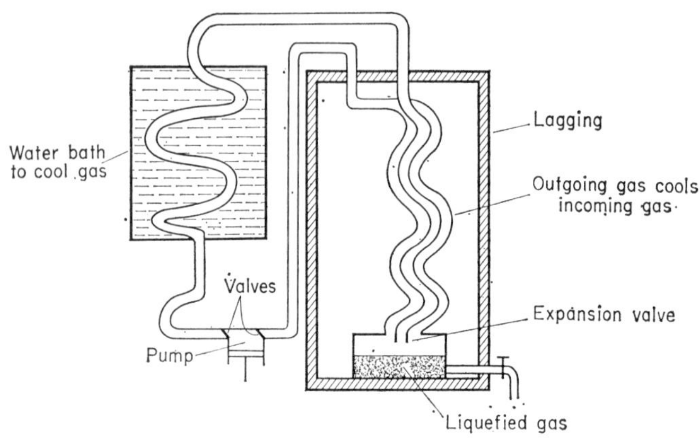

Fig. 148

up and water is used to cool it to room temperature. It is further cooled by cold unliquefied gas returning outside the spiral. The gas is expanded to approximately 1 atmosphere through a valve and a drop in temperature results, due to the Joule–Thomson effect, which is accompanied by partial liquefaction.

(ii) A compressed gas is made to expand in a machine which does work (e.g. an expansion turbine.) The energy for performing this work is withdrawn from the gas molecules and the gas cools.

Specific Heat of a Gas at Constant Volume

The basis of modern methods of measurement is to use a rigid container holding a weighed amount of gas, with an electric coil inside the container to supply the heat. A thermometer, which is usually the coil itself, measures the temperature rise. The thermal capacity of the gas is determined by subtracting the thermal capacity of the container from the total thermal capacity.

For accurate results:

(i) the thermal capacity of the container must be small (this is achieved by constructing it of light material with thin walls);

(ii) heat losses must be avoided (usually done by keeping the surroundings at the same temperature as the gas in the container. A differential thermocouple placed between the container and the surroundings enables the temperature difference to be kept small, either manually or automatically);

(iii) "hot spots" due to hot gases rising must be avoided. A small fan placed inside the container provides the agitation necessary to prevent hot spots.

STUDENT'S NOTES

STUDENT'S NOTES

STUDENT'S NOTES

Index